Telescope Observer's Guide

60 Celestial Objects for Beginners

Contents

Introduction

About this Book

This book has been written for anyone who has a small telescope and who wants to explore the night sky, regardless of their age and experience. My hope is that you'll be inspired to go deeper into the night and discover more of what's "out there."

I'm not really one to use a formal writing style; I'd rather write as though I'm standing beside you and talking to you instead. Likewise, I try to avoid jargon and long pages of explanatory text but, to some extent, that's a little unavoidable. However, if you already own a telescope, know your way around the sky and want to jump right into it, feel free to skip ahead to the star charts on page 39. You don't have to read every word I've written!

If, on the other hand, you're looking to learn a little more (or it's cloudy outside) then feel free to read the other sections too.

In the *Telescope Basics* section, I'll talk a little about the telescopes on the market and which ones might be best suited for you. The *Tips 'n Tricks* section will give you exactly that – a few pointers and handy hints I've picked up over the years. You'll find that many of these are already in practice by amateur astronomers (like yourself!) all over the world.

In *The Things You Can See* I'll discuss the different objects you'll be able to observe (including the planets) and review the information to be found on the individual pages for the objects themselves.

Signposts to the Stars is a short section about the two most important constellations for beginners: Ursa Major and Orion. More specifically, it'll show you how you can use these constellations to find others in the sky. If you're very new to astronomy, you might find this useful.

The *Star Charts & Observing Lists* section is where the fun begins. The section begins with a table; simply look for the current time of year (for example, early November) and then look for the current time of night (for example, 10 p.m.) The table will then tell you which star chart to use. (In this example, that would be Chart 7.)

The chart will show you the night sky as it appears at that time and also includes a list of objects to observe. The list runs from west (setting) to east (rising) but it's not an all-inclusive list as I've focused on the objects that should be highest over the horizon at that time. Some may be too low to make any observation worthwhile.

Next you'll find the biggest section – *The Deep Sky Objects* themselves. Here you'll find information on over sixty "deep sky" objects. For each object you'll find a general star map, a simulated view through a finderscope and a simulation of what you might see through your eyepiece. (See *The Things You Can See* section for more information.)

I always recommend that folks keep a record of their observations (especially if you're new to the hobby) and have included some pages in the *Observation Logs* for you to make some basic notes. There are 20 pages for you to make observations of the deep sky objects contained in this book.

Lastly, there's an *Appendix* that contains some useful supplemental information: for example, object and constellation lists, a basic glossary and recommended resources such as books and software.

Refractors versus Reflectors

There are many different types of telescope but they broadly fall into one of two categories: refractors or reflectors, based on how they gather and focus light (more on that later.)

A refractor is what many people consider to be the classic concept of a telescope. It's basically a tube with a glass lens at one end and an eyepiece at the other. It sits atop a tripod and is usually the kind sold in department and toy stores as a "first telescope" for children.

(If you're looking at a refractor telescope in a department store and it's *not* made by a reputable manufacturer like Orion, I'd walk away. You might think it's a good buy for the price, but it's probably cheap for a reason.)

As you're shopping around, you might come across a telescope that's listed as, for example, a 70mm refractor. This means the lens of the telescope is 70mm in diameter – or just under three inches wide. The size of the lens will dictate how much light is gathered by the telescope and, consequently, how much you'll be able to see as a result.

Again, stay away from department store telescopes that make ridiculous claims like "magnifies up to 1,000x!" In theory, *any telescope* can magnify by 1,000x but the view is worthless because the optics simply won't produce a good quality view. If your optics are imperfect then any imperfections are also magnified by 1,000x.

It's almost a waste of time to try and magnify anything by that much. I suspect a lot of inexperienced folk believe you need a high magnification to see anything – because, after all, the stars and planets are only tiny points of light and those same folks are unaware of everything else you can see at low power.

The Orion Observer 70mm Altazimuth Refractor Telescope.

In reality, only the planets need a high magnification and you can still see a lot at 100x, which is well within the range of a decent small scope made by a reputable manufacturer.

A good refractor is able to produce a relatively high magnification, which makes it an excellent choice for observing the planets. A refractor also has the benefit of being quite portable as, for technical reasons I won't go into here (I promise to restrict the jargon) the size of the lenses typically range from about 60mm to 100mm in diameter.

Orion produces some excellent "first scope" refractors – especially if you're looking for something portable or you're on a budget. In particular, there are some good options for about the price of a family meal, such as the Observer 70mm Altazimuth Refractor, the GoScope II 70mm or the StarBlast 70mm. Each of these are small, light and extremely portable. The GoScope actually comes with a backpack that allows you to easily pack up your scope and take it wherever you want to go – great for family vacations!

The other kind of telescope is a reflector. There are, in fact, many different types of reflectors but I'll try to keep this simple and straightforward. Whereas a refractor works by collecting and focusing the light through a lens, a reflector does the same thing with mirrors.

The light enters through the open end of the telescope tube, called the aperture. It then strikes the large, curved mirror at the bottom of the tube, called the primary mirror, and is reflected back up toward the aperture. Before it can escape again, it hits a much smaller mirror (called the secondary) and is reflected out the side of the tube via the eyepiece.

There's one big advantage to this: a refractor typically requires a long tube because the light isn't reflected and must travel in a straight line. With a reflector, the light is bounced twice – in effect, folded – and so the required tube length is halved. This has the added benefit of allowing for much larger mirrors and, hence, more light gathering power.

There's one more big difference between a refractor and a reflector. A refractor telescope is almost always found on top of a tripod and there are many reflectors that are similarly mounted. However, there's one very popular mount for reflectors that's in very common usage today: the Dobsonian.

A Dobsonian mount is relatively simple. The tube sits upon the mount, which, in turn, sits upon a base on the ground. The mount then turns from side to side and allows the telescope to move up and down.

Dobsonian mounts are designed to carry some pretty large beasts and can be permanently grounded. This makes them a favorite with amateur telescope makers who aren't so concerned with portability but would rather have a powerful telescope instead.

(Having said that, you can also have Dobsonians that are relatively small but are still portable, such as Orion's range of SkyQuest XT scopes.)

Dobsonian telescopes (aka "Dobs") are "light buckets" and are great for deep sky observing but they have one big potential downside – depending on your point of view. Dobs typically aren't motorized.

What does this mean?

Well, when you're looking at an object in the night sky through a telescope, you'll notice that it appears to drift across the field of view as the sky turns above you. Yes, that's right, you will literally see the stars move from east to west.

Having a motorized mount allows your telescope to compensate for this by following, or "tracking," the sky movement; consequently, the object you're observing will stay locked within your field of view. (Without such tracking, objects will drift out of the field of view.) This can be critical if you want to try your hand at astrophotography or want to make a detailed sketch.

The Orion SkyQuest XT6 Dobsonian Reflector

(There are some exceptions to this. For example, Orion produces a range of motorized Dobsonian scopes that will guide you to your object and track it.)

GoTo Scopes

One more thing... technology has evolved a lot over the past few decades and amateur astronomers have certainly benefitted as a result. One of the advancements has been the development of the "GoTo" mount for telescopes.

A GoTo telescope is equipped with a motorized mount that can automatically find objects for you through the use of a computerized hand controller. Before you get too excited, I feel I must tell you this:

Although GoTo's are great – I own one – there's *nothing* that can beat tracking down and "discovering" an object for yourself. A GoTo may be able to place an object in your field of view within seconds, but it's nothing compared to the thrill of using your observational skill (and a decent star chart) to track it down yourself.

So, while all of the objects in this book can easily be found with a GoTo, I *strongly* encourage you to go old school and leave the GoTo behind.

I know what you're thinking though... if I'm not recommending a GoTo, why do I own one myself? Well, there are two main reasons. Firstly, and primarily, it's because of light pollution. I live in the suburbs of Los Angeles, where only the brightest of the bright stars can be seen on any clear night. It's extremely difficult to find any deep sky objects manually under these conditions but the GoTo can find them for me.

The Orion StarSeeker IV 114mm GoTo Telescope

Secondly, when I bought my scope I wanted something with a motorized mount and they're pretty much all GoTo's now. True, you don't have to use that functionality, but when you have a database of thousands of objects at your fingertips, it's kinda hard to resist it.

(Incidentally, I'd argue that whether you're using a GoTo or not, you still need to know what you're looking for. A GoTo mount will often point the telescope very close to the object but it will rarely place it smack in the middle of your field of view. Sometimes, if you're unlucky, you have to slew the scope a little to see your object and if you're not sure what you're looking for, it can be easily missed.)

Which Telescope Should You Buy?

I know I've thrown a lot of information at you and you're probably scratching your head right about now. Don't worry, it gets easier. If you've already bought your telescope, you might even be wondering if you've bought the right one.

In fact, let me address that issue right now. The answer to the question "did I buy the right telescope?" is always "yes. Yes you did." Why? Because anything that allows you to turn your eyes to the stars and to explore the universe is a gift – even if you bought it yourself. When Galileo discovered the four largest moons of Jupiter in 1610 he did it with a simple telescope that makes those department store scopes I warned you about look like the Hubble Space Telescope.

So if you bought your own telescope (even from a department store) and you're wondering if you made the right choice, then don't worry about it. If, on the other hand, you haven't bought one yet (or are looking to buy another – congratulations, you've caught the bug) then here are some recommendations.

Under $125 and/or Portable

You've got some excellent choices for small scopes. If you're looking for something as a beginner scope, maybe for a child or a young adult, I'd go with the Orion SkyScanner 100mm TableTop Reflector. For the price of a family meal, it can provide great views of the Moon, planets and deep sky objects to get you started. (Almost all of the objects in this book should be within reach of this scope, although that will also depend upon your viewing location and your own eyesight.)

From a parent's point of view, if you buy one of these for little Jimmy and he loses interest after a while (I can't imagine why, but who knows) then you haven't lost much. And as an added bonus, the scope is small and very portable. The only downside is that you'll need something like a table to put it on.

The Orion SkyScanner 100mm TableTop Reflector

I've personally bought similar scopes for my nine year old son and my girlfriend's five year old daughter and some of the notes in this book are based upon observations made with a scope this size.

Alternatively, you can go with the Orion GoScope II I mentioned earlier. Although it has a slightly smaller lens (70mm) it's attached to a tripod that can easily be adjusted to suit the observer's height – so there's no need for a table then.

Under $300

If you're more serious, I highly recommend the Orion SkyQuest XT4.5 Dobsonian. This is the telescope I owned for several years and loved it. I spent almost every moonless weekend night outside and was able to "discover" something new every night. I've observed almost every object in this book with that scope and the only reason I didn't observe the rest was because I never got around to it. Many of the notes are based upon my observations during that time.

As the name implies, the XT4.5 is a reflector with a mirror that's four and a half inches wide, or about 114mm. It has a handle and weighs close to eighteen pounds, so it's portable, but you may want a small table to put it on. (I used a folding table I bought from a hardware store for about $20 and that worked just fine.)

The Orion SkyQuest XT4.5 Dobsonian Reflector

The Orion StarSeeker IV 130mm GoTo Telescope

Under $500

If you're looking for a GoTo – I won't judge – then I recommend the Orion StarSeeker IV GoTo 130mm Reflector. The mirror is 130mm wide, which is just slightly more than five inches and every one of the objects in this book should be easily visible with this scope. As an added bonus, it's not too large and heavy that you couldn't potentially take it camping with you or take it outside within a few minutes if the skies suddenly clear.

Again, it's worth repeating – GoTo's are great if you live under light-polluted skies or if you don't have much time to observe – but there's nothing that beats tracking down and bagging a deep sky object yourself.

Tips 'n Tricks

Get to Know the Stars

I can't emphasize this enough. Maybe you're already familiar with the night sky and all the constellations, but if you're not, I'm sorry, but there's no way around this. If you don't know your way around the stars then trying to find some of the hidden treasures will be like starting from a random point in the U.S. and trying to get to Disneyland. One easy way to learn the stars is to use a planisphere, like the one below.

Get a Planisphere

The Orion Star Target planisphere is an excellent choice for anyone wishing to know "what's up" for any given night at any given time.

I've provided some star charts to help you find your way but there's something else you can buy which, I believe, is a huge asset to any astronomer: a planisphere. It's basically a disc with the stars plotted upon it that allows you to "dial up" a view of the sky for any night at any time.

There are two parts to the planisphere: the disc with a map of the entire night sky and a mask that overlays the map and obscures the stars not currently visible.

To simulate the night sky at any time on any date, simply turn the disc so that the desired date is aligned with the time that's printed on the overlaying mask. The planisphere then displays the stars and constellations that are visible at that time.

You're probably thinking "I have an app for that." Yes, you probably do (I have several) – but here's why I don't use the app when I'm observing: firstly, you're relying on your cell phone or tablet battery to keep you going. And if that battery dies and you need your cell phone for the drive home from a dark sky site, you have a potential problem.

Secondly, your eyes need to adjust to the dark (see below) and although many apps will dim the screen, if you hit the wrong button on your device you might find yourself staring at a bright screen instead. And now you've just lost your night vision.

You're probably also saying "how am I supposed to see the planisphere in the dark?" Well, that's what a red flashlight is for (see below.) Orion's Star Target planisphere has you covered from latitudes 30° to 50° North and can be purchased online at www.OrionTelescopes.com

Get a Red Flashlight

A red flashlight is an essential tool for any astronomer. Your eyes need time to adjust to the dark (see below) and if you use a regular flashlight, your night vision will be ruined. On top of that, if you're out with a group and you use a regular flashlight, you run the risk of ruining your fellow astronomers' night vision.

(This also has the potential problem of making your fellow astronomers angry!)

All that being said, your eyes aren't adversely affected by red light and, many moons ago, astronomers would buy some clear plastic red tape to cover their flashlights. Nowadays you don't have to go to all that bother as a red flashlight is pretty inexpensive.

Orion produces several versions, but I've always particularly liked the RedBeam II as it allows you to adjust the brightness of the light. This makes it easier for you to view your planisphere, star charts or even (hopefully!) read this book.

Orion's RedBeam II flashlight allows you to vary the brightness of the red light.

Let Your Eyes Adjust to the Dark

Many people are aware that their eyes grow accustomed to the dark. Stand in a lit bedroom at night, with the curtains drawn and then turn out the lights. What do you see? Probably not much if the room is properly dark.

Likewise, if you need to get up in the middle of the night and you turn the light on, what happens? How do your eyes feel? More than likely you'll be dazzled by the sudden burst of light.

It takes your eyes about an hour to fully adjust to the dark and you don't want to lose that sensitivity (hence the red flashlight.) Without that sensitivity you may not be able to see the finer details in some of the deep sky objects (such as nebulae and galaxies) and some of those objects may not be seen at all.

So what do you do while you're waiting? Well, I for one don't wait around. Take some time to enjoy the night sky. If there's a crescent Moon in the sky, I might take a look at that (but with the use of a filter – see below.) *Don't* observe the Moon after your eyes have adapted to the dark as its brightness will ruin your night vision again.

Look at the planets, if any are visible; although, again, for a bright planet like Venus, a filter may be required to avoid being dazzled. Multiple stars are also a good choice as their light is concentrated in a single, bright point and most are easily seen.

As my eyes adjust, I also take some time to look at the brighter star clusters, such as the Pleiades or the Praesepe. These are often comprised of many bright stars and can be a very welcome distraction as you wait for your night vision to kick in. I *do* also recommend you revisit them later in the session as you'll frequently see many more of the fainter stars once your eyes have adjusted to the dark.

Beware the Moon

The waxing crescent Moon, taken with an Orion SkyQuest XT4.5. Image by the author.

You may notice that I don't talk about the Moon in this book. There's a reason for this. As pretty as it is, the Moon (like an overcast sky) is not my friend. The Moon brightens the sky, just as the Sun does, so instead of the nice, dark sky you'll get on a moonless night, you'll be looking at a deep blue sky instead.

How does this affect your observing? Well, some of the fainter fuzzies (galaxies and nebulae, specifically) will be lost against the glow of the sky. On the plus side, multiple stars are often unaffected as, again, their light is concentrated into a single bright point that can easily shine through.

I don't typically observe when the Moon is out, or, if it's a crescent Moon I'll observe if it's relatively close to the horizon (as a crescent Moon often is.) Consequently, I'll typically wait until about three or four days after the Moon has turned full. By that time, the Moon won't be rising until some time after sunset, giving me the evening hours to conduct my observing. It stays that way until about two or three days after new Moon, when the Moon sets a few hours after the Sun and I can use the rest of the evening for my observing.

Anything between an evening crescent and just after full Moon is a problem because the Moon is visible for much of the evening.

That's not to say the Moon isn't worth observing, because it is – honestly, it can be stunning – and there are a myriad of books and maps out there that will help you with that. But since this book focuses on the "deep sky" objects of stars, star clusters, nebulae and galaxies, I won't be talking about it here. (I *do* recommend you take the time to enjoy the lunar landscape though!)

The Orion Variable Polarizing Filter.

If you're going to look at the Moon, I highly recommend you get yourself a lunar filter for your eyepiece, especially if the Moon is between its half phase and being full. For example, I use Orion's Variable Polarizing Filter. This allows you to adjust the darkness of your view and blocks anything from 60% to 99% of the light coming in. Not only is this great for reducing the Moon's overpowering glare, but it can also help to protect your night vision when observing the brighter planets at low power.

Let Your scope Adjust to the Air Temperature

Just as your eyes need time to adjust, so does your telescope. Admittedly, this is something I often neglect to do as I'm so keen to get out there that I frequently skip this step. Also, I'm easily distracted and will often find myself getting sidetracked when I should be putting the scope outside.

Why is it important? Well, your telescope is probably kept inside your home at a reasonably comfortable temperature. When you take it outside, the air temperature will be different. It might be warmer or it might be cooler, but until your telescope reaches the same temperature you might find the images are not as sharp as they could be.

If you can handle the wait, take your telescope outside about an hour before it's time to observe. Look at it this way – it can take about ninety minutes for the sky to get dark after sunset anyway, so if you take it outside within the first thirty minutes after the sun goes down and *then* go out after twilight has vanished and it's truly dark, then you should be fine.

Locate the Object with Binoculars First

If you've never observed the object before, using a pair of binoculars can make the search easier and also help you to familiarize yourself with that particular patch of the sky. Many of the objects in this book can be seen with binoculars and a standard pair of 10x50's should do the trick. I'll often grab my binoculars when I only have a few objects to observe (or too little time.) I have included my notes on the view through binoculars in the descriptive text for each object.

Orion UltraView 10x50 Binoculars

As always, there are plenty of binocular options available and Orion produces a wide range to choose from. For example, the UltraView 10x50 binoculars will provide excellent views of the sky and can be an effective way to get started in astronomy.

Likewise, many telescope finderscopes will actually magnify the sky a little – often by about 6x. Although not as powerful as binoculars, you should still be able to see almost all the objects in this book within the finder. I've included a graphic depicting the 6x finderscope view beside the text for each object.

Learn to Calculate the Magnification of Your Eyepieces

This is actually very easy but you need to know two important numbers before you begin: the focal length of your telescope and the focal length of the eyepiece you're using.

What's focal length? Well, it's basically the distance that light must travel from entering your telescope (or eyepiece) in order to reach your eye. It's always measured in millimeters and it can always be found on the telescope or eyepiece itself.

On the telescope, this information can be found on a label that's often affixed to the tube, either on the side or close to the focuser where the eyepiece is inserted. It'll probably have the make and model of the telescope and the focal length may be represented with the letters FL.

On an eyepiece, it's printed at the top, almost always on the side and sometimes around the lens of the eyepiece itself.

To calculate the magnification of the eyepiece, you simply divide the telescope's focal length by the focal length of the eyepiece. For example, most telescopes come with a 10mm eyepiece (meaning it has a focal length of 10mm.) So for a telescope that has a focal length of 650mm a 10mm eyepiece will give you a magnification of 65x. (650mm ÷ 10mm = 65)

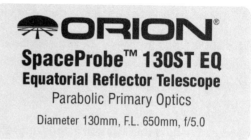

The Orion SpaceProbe 130ST EQ Reflector has a focal length (FL) of 650mm.

Get Yourself a Range of Eyepieces

Obviously then, it's useful to have a small number of eyepieces as this will give you a good range of magnifications to play around with. Most telescopes seem to come with a fairly large eyepiece (for example, a 20mm or 25mm) that will give a low powered view and also a smaller eyepiece (typically a 10mm or a 6mm) that will give a higher powered view.

Personally, I wouldn't go any lower than 6mm because the lens size and field of view typically decrease in relation to the focal length. This means you have to put your eye closer to the eyepiece and, in my experience, this actually makes observing a little uncomfortable. (There's a way around that – it's called a Barlow and I'll talk about that in a moment.)

That being said, if your telescope has a long focal length (say, over 900mm) you can use a 30mm or longer focal length eyepiece to obtain some great wide angle, low power views. (For example, Orion offers a 40mm eyepiece that would produce a magnification of 22½ x on a scope with a focal length of 900mm.)

I'd recommend a 25mm and then adding eyepieces in increments. For example, I own 25mm, 20mm, 15mm and 10mm eyepieces along with a 6mm and a 9mm. (Once you get to 10mm, the sizes tend to decrease in smaller increments.)

You don't have to buy these individually either. Orion offers several eyepiece sets that include a range of sizes. Or, if you prefer something a little more inclusive, look for the telescope accessory kits that include eyepieces, filters and a Barlow lens (see below) which will give you a good start in your new hobby.

My only word of caution is to make sure you get the right barrel size for your telescope. The vast majority of telescopes have a focuser that will fit eyepieces with a barrel 1.25 inches wide, but there are some that will take slightly different sizes. If you're not sure, check your telescope's manual or contact Orion before buying.

But why do you need a range of magnification anyway? Surely a high and a low powered eyepiece is enough? Potentially, yes, but you'll soon learn that when it comes to observing deep sky objects, there comes a point when you can magnify an object *too* much and it loses its aesthetic appeal. A star cluster is prettier when you can see the whole cluster and it's set against a backdrop of faint stars. Multiple stars (especially those of differing colors) are more attractive when they appear close together, rather than highly magnified and far apart.

Orion offers a wide range of Sirius Plössl eyepieces that provide great views of all the objects listed in this book.

Think you only need the two eyepieces that came with your scope? Observe the deep sky objects in this book and then see how you feel!

Know the Maximum Useful Magnification of Your Telescope

Unfortunately, as I mentioned in the Introduction, your telescope is probably not going to produce a good quality image at 1,000x. In fact, I'd be impressed if you can even squeeze 1,000x out of your scope. There are a number of reasons for this, but suffice it to say, every telescope has its maximum useful magnification and it's important to know this before you start observing (or start buying eyepieces that go beyond this limit.)

Quite simply, the theoretical maximum magnification is about fifty times (50x) the aperture of your telescope in inches, or about twice the aperture in millimeters. For example, my Orion SkyQuest XT4.5 Dobsonian had an aperture of 4½ inches so the theoretical maximum magnification was 225x. (4½ inches is slightly more than 114 millimeters, so that would be 228x.)

However, I've found that this isn't always the case. Realistically, I've found the views can be unreliable beyond half the theoretical maximum magnification. Sometimes they're better, but more often they're not. (For example, the only time I've seen the Great Red Spot on Jupiter is when I had the magnification at the theoretical maximum. However, I don't tend to notice much improvement in many faint and fuzzy deep sky objects.)

Therefore I'd go for 25x the aperture of your telescope in inches or – more conveniently – the aperture of your telescope in millimeters. So I'd usually stick to a magnification of 114x or less on my SkyQuest XT4.5 and was never disappointed as a result.

Invest in a Barlow Lens

I've recommended that you buy yourself a range of eyepieces and although you can often buy these for a reasonable price, some can be quite expensive. Also, the smaller the focal length of your eyepiece, the smaller your field of view, which can make it hard to observe.

The solution? A 2x Barlow lens. This is basically a short tube that slots into your telescope's focuser, where the eyepiece normally goes. You then pop your eyepiece into the Barlow and *voila!* Your magnification is instantly doubled. And the best part is your field of view remains the same.

For example, a 20mm eyepiece in my SkyQuest XT4.5 gave me a magnification of about 45x with a nice, wide, comfortable field of view. But if I'm using a 2x Barlow, I double up to 90x and still have the same comfortable field of view.

Orion, of course, has got you covered with their own Shorty 2x Barlow lens. Try this with your telescope and the advantages soon become obvious. You don't just get a higher magnification but a more enjoyable observing experience. As an added bonus, you'll need potentially fewer

The Orion Shorty 2x Barlow Lens

eyepieces so you won't have to carry as much equipment around. It's easier on the eyes *and* the wallet.

Buy a Case for Your Equipment

The Medium Deluxe Accessory Case has enough room to hold all your observing accessories.

If you haven't already, try carrying your eyepieces and equipment out to your scope on a clear, crisp night. Unless you have something to carry them in, you might find yourself making several trips as you carefully move them to your observing location.

As inconvenient as this can be, you also run the risk of dropping and damaging your eyepieces during the move – not the best way to start an evening's observing! Buy yourself a case and protect your equipment (and your investment) by securing them safely in a foam padded environment.

For example, Orion's Medium Deluxe Accessory Case has enough room to store eight standard sized eyepieces and plenty of room to store other essential equipment, such as Barlow lenses, filters and finders.

Start with a Lower Power Eyepiece and Work Your Way Up

When I'm observing, I always start at a low magnification and then increase it. For starters, it's easier to find your target and many star clusters are best observed at low magnification anyway. If the object is easily seen, then I'll gradually increase the power to see how the view changes.

If, however, the target isn't seen, then I'll check to make sure I'm looking in the right place and then up the power to something closer to 100x. This is also true of multiple stars that can't be split at low magnification. For example, I may only see a single, bright star at 35x so I'll up the power to around 100x and see if my luck improves. If the star's split, I'll start decreasing the magnification to find the point where it becomes a single star again. Quite frequently, I've gotten down to 35x again and still been able to split the star – but only because I now know where to look for the companion.

Focus on the Stars

If you're staring at something faint and fuzzy (for example, a globular star cluster) you'll often find it difficult to properly focus the view. Instead of trying to focus on the object, focus on one of the nearby stars instead. And if you don't see any in the same field of view, try moving the scope slightly so that some are visible. It's far easier to focus on a single point of light than something grey, misty and ill-defined!

The Colors of Multiple Stars

I've found that star colors can be very subjective and may even seem to vary from night to night. There are a number of reasons for this. Unless the colors are quite strong (such as the famous double star Albireo), what you see might vary depending upon the sky conditions, the equipment you're using, the height of the star in the sky and, frankly, your own eyesight.

So if I've described a blue-white star and you're seeing just white – or maybe even pale gold or yellow – then there's nothing wrong. It just means that, like almost everything in life, you're having a slightly different experience. (If you read the observations of others, you'll often find a wide range of colors being described.)

If you're having difficulty seeing any color at all, try de-focusing the view. As the star loses clarity, it will appear larger, spreading the light over a wider area in your field of view and sometimes making the color easier to see.

Use Averted Vision

What's averted vision? Basically, it's observing a little off to the side of an object, rather than directly at it. Without getting too technical, your peripheral vision is more sensitive to light (and movement) than your direct vision. It's thought this enabled our ancestors to detect potential nearby predators before they became a more immediate threat.

If you can't see an object, try using averted vision. If that doesn't work, try tapping the telescope tube gently to make the view move, as averted vision is also good for detecting movement.

Once you've found it, looking at the object with averted vision can reveal detail not visible with direct observation. However, there's a trade-off.

From personal experience, it seems that while averted vision may help you to see a faint object, it's usually lacking in color, whereas direct vision may show shades of green or blue. The center of your eye is better suited for providing a good, all-round color view while it seems the edge of your vision is best suited for detection and motion.

Keep an Astronomical Journal

One of my journals, complete with chicken-scratch handwriting. Image by the author.

As a kid I used to keep one and wrote an entry for every object I observed on every night. Well, almost. I admit to skipping it sometimes. There's a couple of good reasons for this. Firstly, it's a great way to keep track of which objects you've seen and to feel a real sense of accomplishment as your tally increases.

Secondly, I use it to plan my session. Before I go outside, I'll make a list of all the objects I want to observe, along with any additional information I'll need – for example, the object's catalogue number or coordinates. As I'm going outside, I'll also make a note of the weather, sunset time and Moon information such as its phase, illumination and rise or set time.

Lastly, it's simply a good reference tool. It's a rare object that only gets observed once and you'll return to many of your favorite objects over and over again. However, I don't re-read my notes before I observe that object again. Why? Because it can actually prejudice your viewing experience and influence your expectations.

I've also found it sometimes spoils the experience a little. There's nothing better than observing an object for the first time and discovering something new, and when you're starting out there's literally hundreds of objects to discover. If you don't review your notes beforehand, it's easy to forget what you've seen and then you go to the eyepiece and rediscover that object all over again. (There's an exception to this rule. Sometimes I'll observe an object but make a separate note to re-observe it later, looking for a specific detail.)

Similarly, unless you're having difficulty observing an object, try to avoid the notes of others beforehand. I know this sounds contrary to the concept of this book, but I'm hoping the text will give you an idea of what to expect while encouraging you to have your own personal experience. And that's really the point of astronomy. It's a personal experience for everyone. Reading someone else's opinion of an object beforehand can color that experience and, I feel, you lose something as a result.

Compare your notes after you're done for the night, not before. That way you can find out if that star was truly multiple or if you were just imagining it. Or you'll read about a detail you missed and you'll want to go back and observe that object again.

To start you off, I've included 60 observation log templates (starting on page 133) so you really have no excuse! Go, observe, keep a journal and compare notes later. You'll enjoy it far more that way.

The Things You Can See

About the Deep Sky Objects

There are basically five types of objects you can observe with your telescope:

* Solar System Objects
* Stars
* Star Clusters
* Nebulae
* Galaxies

All of the objects listed in this book should be easily seen with a small telescope. By small, I mean something with an aperture of 70mm (about three inches) to 125mm (about five inches.) Specifically, I've tried to list objects that can be seen with a small 70mm scope, but it'll be easier if your scope is larger.

All of the objects are either easily seen with just your eyes (making them very easy to find) or else they lie very close to a bright star. That way, you'll be able to easily locate them by first finding the star itself. The object will always be within the same finderscope field of view as the star.

Please be aware of one thing: although I've seen the vast majority of these objects from the suburbs of Los Angeles (hello, light pollution) and frequently while they're setting in the west (as opposed to high overhead, where the air is clearest) whether you can see the object will greatly depend upon your location, your equipment and your own eyesight. I've chosen objects that should be visible regardless, but I'm not you, I don't know your personal circumstances and, consequently, I can't guarantee your own personal experiences.

That being said, I'm confident you'll be able to see the vast majority, if not all of them. (A word to readers in the United Kingdom: there are two objects you won't be able to see from your latitude and I apologize in advance for that. These are the Butterfly Cluster and Messier 7. Unfortunately, they're too far south to rise over the horizon and I feel bad about that because I'm English myself. If it's any consolation, they barely skim the southern horizon from Los Angeles while observers in the southern hemisphere are lucky enough to see them pass overhead.)

Before we get started, let's take a few moments to review the information to be found on the individual pages for the objects.

At the top you'll find the object name or identifier. The names of the stars are often Arabic in origin, but you'll notice some may have names like "Gamma Delphini." The first part of the name (gamma, in this example) is a letter of the Greek alphabet. Thousands of years ago, the Greeks assigned letters to the stars based upon their brightness. So the brightest star in the constellation would be assigned the letter Alpha – their equivalent of the letter A. The second brightest would be Beta, then Gamma, Delta and so forth. (There's a table detailing the Greek alphabet in the Appendix on page 162.)

The second part of the name refers to the constellation the star belongs to – in this example, Delphini refers to Delphinus, the Dolphin. (Again, there's a list of constellations in the Appendix on page 165)

The system has been refined and updated since then, but it's still very widely used. After all, not every star in the sky can have a name and this is an easy, convenient alternative.

Some of the other objects (the star clusters, nebulae and galaxies) may have names, such as the Andromeda Galaxy or the Owl Cluster. More often than not, they're named after whatever the object resembles (such as an owl) or sometimes the constellation it resides in (such as Andromeda.)

The book also mentions Messier objects. These are objects that were noted by the French comet hunter Charles Messier in the 18th century. Some deep sky objects can appear distinctly comet-like (especially globular star clusters) and Messier wanted to avoid confusing them with any potential discovery.

There are 110 Messier objects in all and this book contains 22 of them. Some are actually quite clearly not comets – for example, the Pleiades – and many astronomers have wondered why Messier went to the trouble of cataloging them. Other objects, such as the famous Double Cluster or the Owl Cluster, are easily visible and yet Messier didn't list them at all.

One other catalog is also well known and well traversed by astronomers – the New General Catalog, or NGC. This list was compiled in the 19th century and contains nearly 8,000 deep sky objects. However, many of them may be beyond the reach of a small scope. Incidentally, all of the Messier objects were also assigned an NGC number.

Charles Messier, circa 1770. Public domain image.

On one side of the object page you'll see three graphics. The first is a general map of the area where the object may be found. The image spans 45° of sky and in the center is a circle. This circle denotes the area of sky seen through a 6x finderscope, as depicted in the second image.

The finderscope image is right-side up; in other words, north is at the top of the picture and south is at the bottom. I mention this because some finders will flip the view; please be aware of this as you're looking for your object. Both these images were created using the *Mobile Observatory* app by Wolfgang Zima and used with permission.

The third image on the page attempts to depict the view through your eyepiece. I say *attempts* because I don't know exactly what kind of eyepiece you're using or the field of view you'll get, but this should be a pretty good representation. You'll see the magnification in the top left corner and in the bottom left is an indication of north and east. Your telescope will flip the view; it may just be vertically, or horizontally or both but the view depicted should match many mainstream telescopes.

There are a few other things to bear in mind when looking at these eyepiece views.

With the exception of open clusters, the other deep sky objects are drawn depicting their true size in the sky. In reality, you won't see the object this large in your finder or eyepiece unless you're observing under very dark skies with a large telescope. As stated before, what you see will depend upon your equipment, your location, the sky conditions at the time and your own eyesight. The same is true of some of the background stars depicted; how many you'll see will, again, depend upon those same factors.

Also, it's difficult to accurately depict the multiple stars because, in order to show which stars are brightest, it's necessary to make the dots representing the stars larger. For the sake of clarity, some of the multiple star images may have large magnifications (e.g., 217x) in the corner. In reality, you can probably get a similar view at about half that magnification.

The eyepiece views were created using *Sky Tools 3*, an excellent program for your PC or Mac that allows you to plan an evening under the stars. As you can see, it also accurately simulates the view through your eyepiece. Like *Mobile Observatory*, I consider it invaluable. (See www.skyhound.com for more information.)

My thanks to Greg Crinklaw for his permission to use these images.

The Eagle Nebula in the constellation of Scutum. Credit: ESO (European Southern Observatory.)

You'll notice I don't have any photos of the objects and there's one very good reason for this: more often than not, they don't accurately depict what you'll see through the eyepiece, whereas *Sky Tools* does. Too many new astronomers get discouraged because they see beautiful, colorful images in magazines and online and expect to see something similar. Those photographs are created by dedicated individuals who'll spend hours combining many images and fine tuning the final result until it's the best it can be.

I hate to break it to you, but your telescope is not called Hubble either. I don't want you to be discouraged; the night sky can be stunning and awe inspiring but sometimes you have to use a little imagination and truly take into account what you're looking at. I've tried to do that in the text that accompanies each object.

On the other side of the page is some information about the object. You'll see its designation (for example, Gamma Andromedae, Messier 45 etc.), the constellation it belongs to and then we have two coordinates, R.A. and Declination.

R.A. is short for Right Ascension and this, coupled with Declination, will give you the exact position of the object in the night sky. Think of it this way – if you were to plot the stars on a globe, then R.A. and Declination would be the sky equivalent of longitude and latitude, respectively.

I debated whether to include them as the objects are easily found once you get to know the constellations. However, some of you may be using a star chart, such as the *Pocket Sky Atlas* by Roger W. Sinnott, and anyone with a GoTo can enter the coordinates into their hand controller and the telescope will find it for you. (Most of the objects will be included in the GoTo's database anyway.)

Below the coordinates you see what type of object it is – for example, multiple stars and open clusters are the most common. I'll talk about these in a moment.

Next you'll see two sets of stars – one for Location and one as a general Rating. In terms of Location, an object with three stars means it's easily visible to the unaided eye. So, for example, a bright star will have three stars for Location.

Objects with two stars for Location are those that can be found by placing a nearby bright star in the finder and the object will appear in the middle. Objects with one star for the Location are those that may be particularly faint or might require you to have the bright star on the edge of the field of view and the object will appear on the opposite edge.

The general rating is more straightforward. I've given a three star rating to the objects I consider to be unmissable and those I always take time to observe. Objects with two stars still provide something worth seeing while those with one star may be appear faint and best observed under clear, dark skies. (As opposed to the light polluted skies of suburbia.)

There are a lot of objects in astronomy that may seem like a "one star" object at first glance, but remember this: there's a sense of accomplishment that comes from being able to track it down and, as you observe it, consider what you're looking at. Again, I've provided some notes to help with this.

(Please also bear in mind that I want you to have realistic expectations of what you'll see. Astronomy is awe-inspiring, challenging and fires the imagination.)

Lastly, I've indicated when the object is best seen. You might be able to see the object at other times of the year but this will sometimes require you to observe during the early hours of the morning. So the seasons listed here indicate when the object is at its best visibility in the evening. (This is also why I included the star charts earlier in the book – so you can see what's up in the evening and immediately start observing.)

Below that you'll see some text about the object. This information is predominantly based upon my own personal notes and experiences, both while observing from Oklahoma and also from Los Angeles. Again, my goal was to provide you with a realistic idea of what you might see from either a suburban or a city location. I've read a lot of books written by astronomers lucky enough to observe under clear, dark skies but I understand that this isn't an option for everyone.

Solar System Objects

Technically, objects within the solar system aren't deep sky objects but I wanted to briefly discuss them as they're well worth taking some time to truly appreciate them. As you might expect, there are many other books on the market that discuss observing the planets and one of the best is *The Planet Observer's Handbook* by Fred W. Price.

The five brightest planets – Mercury, Venus, Mars, Jupiter and Saturn – are visible with just your eyes (although Mercury can be tricky) with Venus, Jupiter and Saturn providing particularly good views through a small telescope. With a little more experience, you'll also be able to spot Uranus and Neptune.

Before you do anything, you'll need to know if any planets are currently visible and this is where the *Mobile Observatory* or Orion's *StarSeek* apps definitely prove their worth. Alternatively, you can try the *Sky Tools 3* observation session planner program or one of the more basic planetarium programs, such as *Stellarium* (a free version is available for PC's and Mac's.)

Mercury and Venus are both known as *inner planets* because they're closer to the Sun than the Earth. Consequently, they can never appear opposite the Sun in the sky (unlike the other planets) and can only be seen in the early morning or early evening sky. So, for example, if you're an early riser you might catch them both in the pre-dawn twilight, but if you're looking for them in the evening, you can catch them in the twilight after sunset.

When you're checking their visibility, you'll need to find out when they reach their greatest elongation from the Sun. This is when the planet appears furthest from the Sun in the sky and is well-placed for observation.

Here's where it gets slightly confusing. If the planet is at greatest *eastern* elongation it's visible in the evening sky, but in the west. If it's at greatest *western* elongation, it actually means it's visible in the pre-dawn sky, in the east.

There's a table in the Appendix that lists the elongation dates between 2016 and 2025 (see page 163 for details.)

You might need some help finding Mercury as it's often low on the horizon and can be a little faint against the twilight sky. It often appears pink-white and may be twinkling rapidly. Using binoculars will definitely help you. Mercury, like the God it is named after, is fleet-footed and may only be visible for a few weeks at a time.

Venus is a lot easier. In fact, Venus is unmissable. It appears as a brilliant white star, sometimes quite high above the horizon, often for hours after sunset and for months at a time. It's been mistaken for a UFO and provides a beautiful addition to the evening night sky, especially during the holiday season. After the Sun and Moon, it's the third brightest object in the sky.

Through a telescope both planets will show phases like the Moon. Venus is closest and will therefore appear largest with Mercury appearing much smaller. You'll only need about 35x to clearly see the white disc of Venus while about 100x will provide a great view.

The planet Venus, as imaged by Lukáš Kalista. Image credit: Lukáš Kalista.

Mercury's disc is problematic at low power – you'll probably be able to glimpse it at around 35x but you won't be able to clearly see it until you get to around 80x or 90x. To me, it's always appeared a tan or dull gold color.

All the other planets orbit further away from the Sun and can appear opposite the Sun in the sky. When this happens, the planet is said to be at *opposition* and it's the best time to observe the planet as it will rise at sunset and set at sunrise. It's therefore visible all night and will appear at its brightest and largest.

As with Mercury and Venus, there's a table in the Appendix that lists the opposition dates for Mars, Jupiter and Saturn between 2016 and 2025 (see page 164 for details).

Unfortunately, Mars only reaches opposition once every couple of years and isn't always conveniently placed for observation. Also, the size of its disc will vary, depending upon how close the planet is to us at the time. To the unaided eye it appears bright and coppery and a small telescope will show a tiny disc at low magnification.

At its best (and with higher magnification) you'll be able to see tiny dark markings on its surface and maybe even the glint of an ice cap at the poles. I've seen markings at 54x but you might need to go higher. A magnification of about 100x seems fairly reliable.

Mars during one of its closest oppositions to the Earth. Image credit: Rochus Hess.

Jupiter and Saturn present, by far, the best views of any planet in the solar system and I promise you will never get bored of viewing them.

Many people are wowed by Saturn when they see it for the first time. Personally, I love Jupiter. Why? Two reasons: firstly, because you get to see more and secondly, you can start an observing session with Jupiter and then come back to it just a few short hours later and you'll see that things have changed.

For example, even a pair of steadily held binoculars will reveal the planet's four largest moons – Io, Europa, Ganymede and Callisto. These moons are collectively known as the Galilean satellites after the Italian astronomer Galileo Galilei who first discovered them in January 1610. These moons orbit the planet fairly quickly and can be seen to cast shadows against Jupiter's disc and disappear in eclipses behind the planet.

Beyond these four moons you'll also see dark stripes across the disc of the planet. Typically you'll see two (the northern and southern equatorial belts) at around 40x and if you increase the magnification to around 70x or 80x you may also notice dark patches at the poles.

You probably won't see the Great Red Spot. This is the Earth-sized storm that looks like Jupiter's eye staring out into space in many of the photographs of the planet. Despite many years of looking, I was only able to see it through a 130mm reflector with a blue eyepiece filter to help bring out the contrast in the planet's features. Even then, if I hadn't known it was currently visible, I probably would have missed it!

Saturn, of course, is famous for its rings and is usually the planet that beginners want to see first. There's something about a person's first view of the planet that often stays with them and many amateur astronomers can clearly recall when they saw the planet for the first time.

The giant planet Jupiter. Image credit: Rochus Hess.

You've probably seen a lot of photos of the planet but when you actually see it for yourself, you'll notice a third dimension that you just can't experience any other way. It looks like a Christmas tree ornament, impossibly hanging there in space. It's easy to understand how the first telescopic observers, some four hundred years ago, were completely confused and bewildered by the rings. Nowadays, thanks to missions like the *Voyager* and *Cassini* space probes, we know the rings to be comprised of millions of chunks of ice, some no bigger than a fist, others as large as a house, all orbiting Saturn in unison.

The planet Saturn, as imaged by Lukáš Kalista. Image credit: Lukáš Kalista.

Like Jupiter, you'll also be able to see a few moons, most notably Titan, the planet's largest. But unlike the moons of Jupiter, they don't move quickly and the view won't considerably change within a few short hours. However, come back the following night and you'll see some differences.

Beyond that, you may see a gap in the rings (called Cassini's Division, named after the astronomer who discovered it) and maybe some faint markings on the disc, but that's about it. To be honest, for me it's like a childhood friend I've grown apart from. I'll still stop by and visit if it's around, but I don't usually stay for long and I don't find there's much to talk about!

Uranus and Neptune can also be seen but you need to know where to look for them. At low power (say, around 35x) they only appear as tiny, starlike points so even if you're using a GoTo, you'll need to know which are the stars and which is the planet. Having said that, there *is* a way you can tell. Both planets will show a fairly distinct color – a pale turquoise tint for Uranus and a sky blue hue for Neptune.

Uranus might also show a disc; I've possibly seen it at 26x but I've typically needed at least 50x to be sure. Neptune is trickier and I've only seen a very tiny disc at a magnification of 160x or higher.

The same is true for asteroids. You'll need to know exactly what you're looking for and be familiar with the background stars. Unfortunately, unlike the major planets, they're too small to appear as anything but a tiny star-like point in your telescope and can often be disappointing to new astronomers.

One last thing: comets. Occasionally, a reasonably bright one might come close and make itself visible to amateur astronomers with small telescopes. How bright it will be and how it will appear in your scope really depends on the comet itself.

No two comets are alike and none is guaranteed to be a spectacular celestial event. I've observed a number and they typically look like misty spheres with a star-like point (the core of the comet) at its center. Some comets will also show a tail while others might have a greenish tint to them.

Be sure to join an astronomical group (be it online or in person) to keep up-to-date with any potential surprise visitors that might be coming around.

Comet Hale-Bopp in 1997. Image credit: Franz Haar.

Multiple Stars

The double star Albireo, aka Beta Cygni (see page 125.) Image credit: Henryk Kowalewski.

I hate the phrase "star gazing." For me, it always conjures up an image of someone standing still, staring up at the night sky. Or maybe peering through a telescope at a single bright star. Someone else might walk by. "What are you looking at?" they ask. The observer points up at the sky. "That bright star. That one, right there."

Many non-astronomers think this is what "star gazers" do. It's like when I stopped eating white and red meat during the 1990's. People thought I just ate peas and carrots instead. (Incidentally, I've since returned to the dark side.)

Nothing could be further from the truth, especially given that many stars are not single stars at all. In fact, most of them are *multiple* stars. To the unaided eye, they'll appear to be a single star, but when you observe them with binoculars or a telescope, that star is split in two. Some stars may have three or four components.

There are two kinds of multiple stars: those that are true multiple star systems, where the stars orbit one another, and those that are *optical* doubles and only appear close together due to a chance alignment. In reality, they may be light years apart.

Some multiples have components that are equally bright (such as Kuma) while others show stunning colors (such as the famous Albireo.)

The best thing about multiple stars (beside their abundancy and variety) is that they're largely unaffected by moonlight and light pollution. As long as both components are reasonably bright, they'll still shine through the brightened sky caused by an intrusive Moon.

Star Clusters

Messier 6, the Butterfly Cluster in the constellation of Scorpius (see page 113.) Image credit: Ole Nielsen

Star clusters also fall into two categories: open star clusters and globular star clusters. An open star cluster contains tens or hundreds of stars, all of which literally appear clustered together in the same relatively small area of sky. This is not a chance alignment – these stars are genuinely grouped close together in space and are born from the same nebula. Consequently, they're usually quite young and are quite literally, stellar siblings.

Some clusters appear fairly large and only require a low magnification with member stars scattered across the field of view. The famous Pleiades and Praesepe open clusters are excellent examples. Others are small and compact and may require a higher magnification to be properly appreciated. Messier 103 is one example.

The other type of cluster is the globular cluster. These are spherical balls of stars in space and contain thousands of stars within a tightly packed area. They lie thousands of light years away, usually close to the hub of the galaxy, and are very old in comparison to the younger open clusters. It's not unusual for a globular to be over ten billion years old – nearly as old as the universe itself.

Through a small telescope a globular can often appear like the head of a comet without the tail. Alternatively, you might think of it as being a faint and fuzzy star. But increase the magnification and you may be able to resolve some of the individual stars around the outer edges. You may also notice chains of stars or the cluster may appear to be misshapen. Not all globulars are the same! For my money, the best globular in the northern hemisphere is the Keystone Cluster (Messier 13) but Messier 22 is a fine example too.

Whether you're observing open or globular star clusters, you'll get the best views away from the lights but you can still get some great views from the suburbs, even with light pollution.

Messier 13, the Keystone Cluster in the constellation of Hercules (see page 115.) Image credit: Peter Linde

Nebulae

There are, of course, several categories of nebulae but the most common are clouds of gas and dust in space. These are the birthplaces of stars and can cover an area light years in diameter. The Orion Nebula is the best example of this in the northern hemisphere. It's easily seen with the unaided eye as a tiny, misty patch in the sword of Orion and can be a stunning sight in a telescope.

Unfortunately, that's the only example of that kind in the book as most nebulae are fainter and don't appear close to a bright star. However, there are a few others that might draw your attention.

Messier 42, the Orion Nebula in the constellation of Orion (see page 85.) Image credit: Sylvain Billot

One is the Ring Nebula, an outstanding example of a planetary nebula. These nebulae are small and appear disk-like when observed through a telescope, almost like a planet. The Ring is the best and brightest example and literally looks like a tiny smoke ring in space. It's located in the constellation of Lyra and is best seen in the summer.

The other kind of nebula is a supernova remnant. There are only a few of these and only one is easily seen in a small telescope. This is the Crab Nebula and it's the remains of a star that exploded nearly a thousand years ago. The Crab is located in Taurus, the Bull, and is best seen in the autumn and winter.

Galaxies

Messier 31, the Andromeda Galaxy in the constellation of Andromeda (see page 69.) Image credit: Adam Evans

Galaxies come in all different kinds of shapes and sizes but may be disappointing to the beginner. There are a lot of images around showing star-studded spirals in space, but the reality is that you're likely to only see a small, misty patch. Depending upon the object, your equipment and location, you may be able to see some shape and structure and patience definitely pays dividends, but don't get your hopes up if you're affected by light pollution.

Also, the vast majority of galaxies are small, faint and can be difficult to locate. With experience, you'll be able to spot a number of them, even with a small scope, but to begin with there's only one that's easily seen. The Andromeda Galaxy appears as a misty patch with the unaided eye and is conveniently located close to a number of bright stars.

(I did also consider including another, the Triangulum Galaxy, but this is harder to locate and can be tricky to see against the background sky.)

Ursa Major

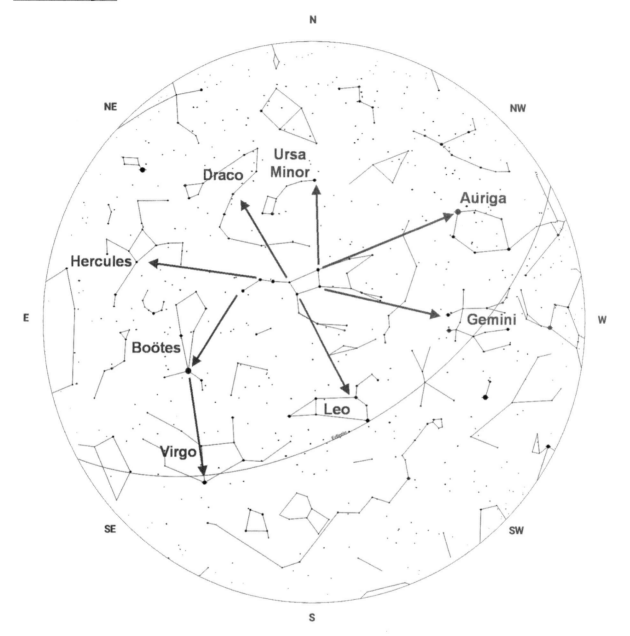

You can use the brightest stars in Ursa Major, commonly known as the Big Dipper (or the Plough in the United Kingdom) to find a number of other constellations. The table below lists the objects to be seen in those constellations and the map above is similar to star chart #12.

Object	Page
Messier 37 (Auriga)	88
Messier 35 (Gemini)	89
Castor (Gemini)	90
Polaris (Ursa Minor)	80
Regulus (Leo)	99

Object	Page
Adhafera (Leo)	100
Algieba (Leo)	101
Denebola (Leo)	102
Arcturus (Boötes)	106

Object	Page
Delta Boötis (Boötes)	107
Keystone Cluster (Hercules)	115
Rho Herculis (Hercules)	116
Kuma (Draco)	117

You can use Orion to find a number of other constellations. The table below lists the objects to be seen in those constellations and the map above is similar to star chart #6.

Object	Page
Pleiades (Taurus)	81
Crab Nebula (Taurus)	82
Messier 37 (Auriga)	88

Object	Page
Messier 35 (Gemini)	89
Castor (Gemini)	90
Procyon (Canis Minor)	94

Object	Page
Sirius (Canis Major)	92
Messier 41 (Canis Major)	93

Star Charts & Observing Lists

Star Chart Table

The table on the opposite page shows which numbered star chart should be used at a given time in a given month. These star charts are designed for locations at a latitude of approximately 40° north of the equator. (You can download printable star charts designed for locations at 30°, 40°, and 50° north latitude from www.OrionTelescopes.com/skycharts)

Many astronomical books only have twelve charts (one per month,) or even just one for each season. I've provided twenty-four for greater accuracy and ease-of-use. This gives us a star chart for every hour in the day, rather than a just one for every two hours. (A lot can happen in two hours – for example, bright stars and deep sky objects can rise and set. Go outside at 8pm and then return at 10pm to see how much your view of the sky has changed.)

Please note that if you're observing during daylight saving time, you'll first need to subtract one hour and then refer to the corresponding star chart number. For example, for 10pm daylight time in early August, use chart 9.

Each chart depicts the entire night sky with the outside circle representing the horizon. Around the edges of the circle you'll see the cardinal points (for example, N for North, NE for Northeast, etc.) while in the center you might notice a Z – this stands for Zenith, the point directly above your head.

To use the chart, take the book outside and turn it so that it's oriented to the same direction as the direction in which you're looking. For example, if you're looking toward the north, turn the book so that N is at the bottom. The view in the sky will then match the view depicted in the chart. You can use a red flashlight (see page 14) to clearly see the sky charts in the dark without ruining your night vision.

You'll also notice a curved line running horizontally across the chart. This line represents the ecliptic, which is the path the Sun, Moon and planets appear to take as they move through the sky. You might also notice the constellations that they move through. These are the signs of the zodiac: Aries, Taurus, Gemini, Cancer, Leo, Virgo, Libra, Scorpius, Sagittarius, Capricornus, Aquarius and Pisces. Although twelve are officially recognized, in reality Scorpius is barely touched with a thirteenth, Ophiuchus, playing a larger part.

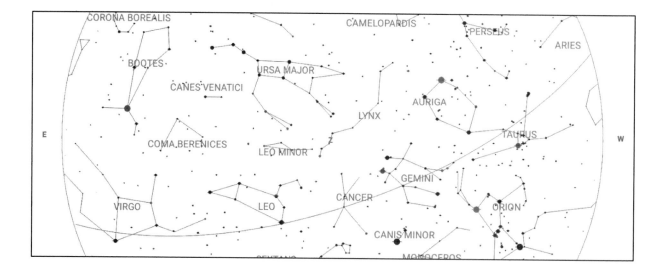

	6pm	7pm	8pm	9pm	10pm	11pm	12am	1am	2am
Early January	1	2	3	4	5	6	7	8	9
Late January	2	3	4	5	6	7	8	9	10
Early February	3	4	5	6	7	8	9	10	11
Late February	4	5	6	7	8	9	10	11	12
Early March	5	6	7	8	9	10	11	12	13
Late March	6	7	8	9	10	11	12	13	14
Early April	7	8	9	10	11	12	13	14	15
Late April	8	9	10	11	12	13	14	15	16
Early May	9	10	11	12	13	14	15	16	17
Late May	10	11	12	13	14	15	16	17	18
Early June	11	12	13	14	15	16	17	18	19
Late June	12	13	14	15	16	17	18	19	20
Early July	13	14	15	16	17	18	19	20	21
Late July	14	15	16	17	18	19	20	21	22
Early August	15	16	17	18	19	20	21	22	23
Late August	16	17	18	19	20	21	22	23	24
Early September	17	18	19	20	21	22	23	24	1
Late September	18	19	20	21	22	23	24	1	2
Early October	19	20	21	22	23	24	1	2	3
Late October	20	21	22	23	24	1	2	3	4
Early November	21	22	23	24	1	2	3	4	5
Late November	22	23	24	1	2	3	4	5	6
Early December	23	24	1	2	3	4	5	6	7
Late December	24	1	2	3	4	5	6	7	8

Chart 1

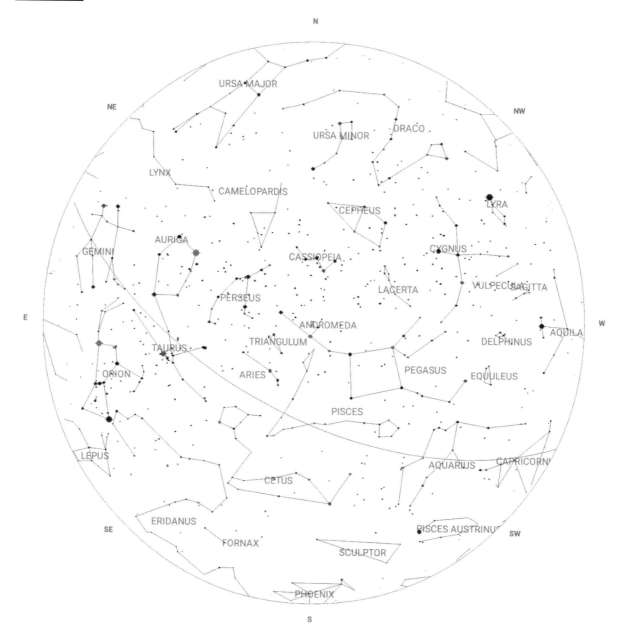

NB: Not all visible objects may be listed. Objects are listed based upon their visibility, from west to east.

Object	Page
Kuma	117
Double Double	119
Sheliak	120
Albireo	125
Messier 29	126
Zeta Sagittae	127
Gamma Delphini	130
Messier 15	131

Object	Page
Pi Andromedae	68
Andromeda Galaxy	69
NGC 752	70
Almach	71
Achird	72
Owl Cluster	73
Messier 103	74
NGC 663	75

Object	Page
Double Cluster	78
Messier 34	79
Mesarthim	76
Lambda Arietis	77
Pleiades	81
Messier 37	88
Messier 35	89
Meissa	84

Chart 2

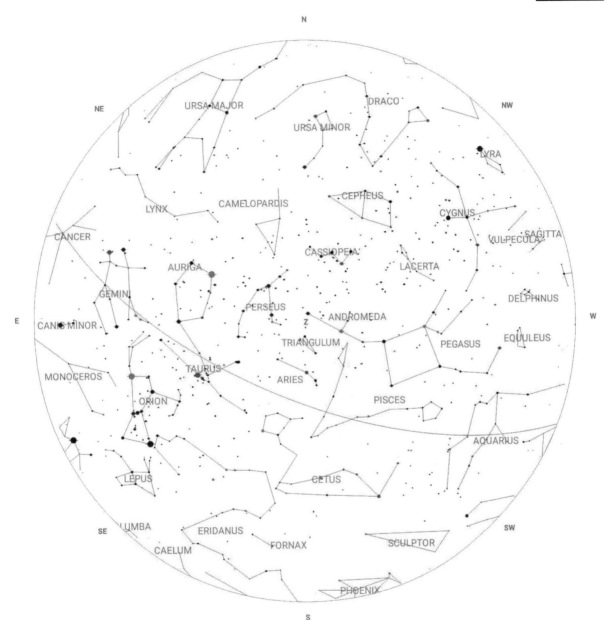

NB: Not all visible objects may be listed. Objects are listed based upon their visibility, from west to east.

Object	Page
Messier 29	126
Messier 15	131
Pi Andromedae	68
Andromeda Galaxy	69
NGC 752	70
Almach	71
Achird	72
Owl Cluster	73

Object	Page
Messier 103	74
NGC 663	75
Double Cluster	78
Messier 34	79
Mesarthim	76
Lambda Arietis	77
Pleiades	81
Crab Nebula	82

Object	Page
Messier 37	88
Messier 35	89
Castor	90
Meissa	84
Mintaka	83
Sigma Orionis	86
Orion Nebula	85
Polaris	80

Chart 3

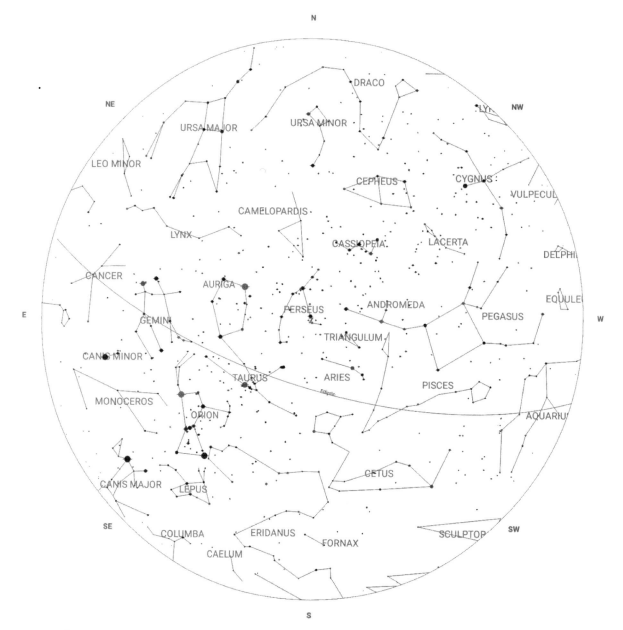

NB: Not all visible objects may be listed. Objects are listed based upon their visibility, from west to east.

Object	Page
Pi Andromedae	68
Andromeda Galaxy	69
NGC 752	70
Almach	71
Achird	72
Owl Cluster	73
Messier 103	74
NGC 663	75

Object	Page
Double Cluster	78
Messier 34	79
Mesarthim	76
Lambda Arietis	77
Pleiades	81
Crab Nebula	82
Messier 37	88
Messier 35	89

Object	Page
Castor	90
Meissa	84
Mintaka	83
Sigma Orionis	86
Orion Nebula	85
Gamma Leporis	87
Beta Monocerotis	91
Polaris	80

Chart 4

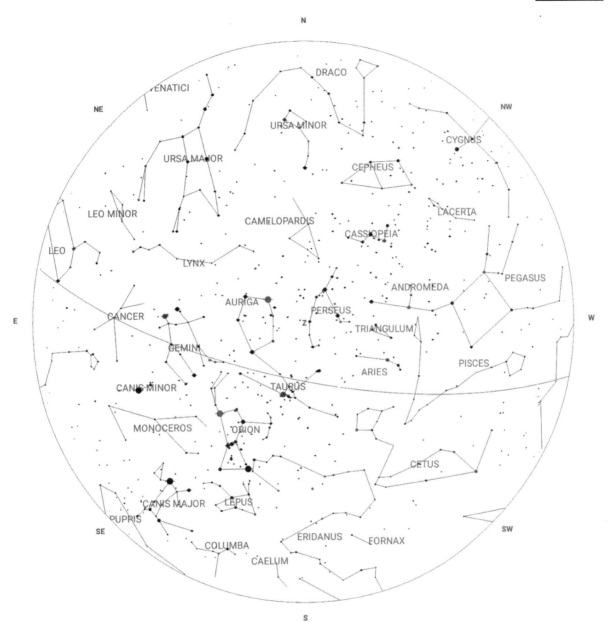

NB: Not all visible objects may be listed. Objects are listed based upon their visibility, from west to east.

Object	Page	Object	Page	Object	Page
Pi Andromedae	68	Double Cluster	78	Castor	90
Andromeda Galaxy	69	Messier 34	79	Meissa	84
NGC 752	70	Mesarthim	76	Mintaka	83
Almach	71	Lambda Arietis	77	Sigma Orionis	86
Achird	72	Pleiades	81	Orion Nebula	85
Owl Cluster	73	Crab Nebula	82	Gamma Leporis	87
Messier 103	74	Messier 37	88	Beta Monocerotis	91
NGC 663	75	Messier 35	89	Procyon	94

Chart 5

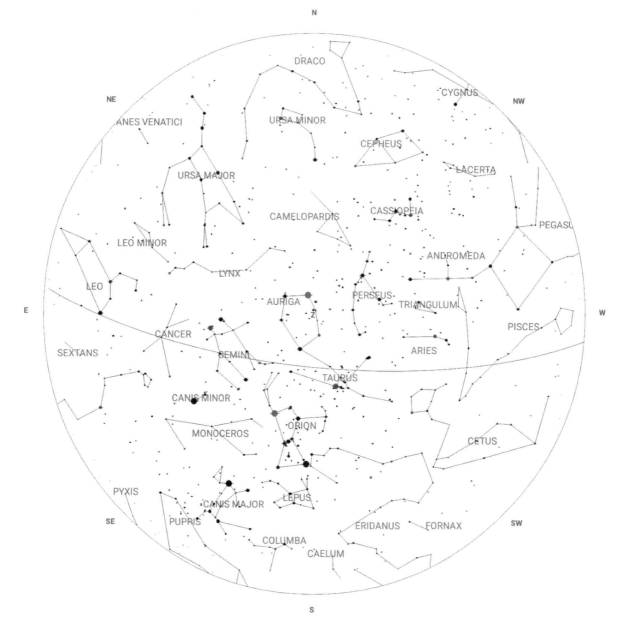

NB: Not all visible objects may be listed. Objects are listed based upon their visibility, from west to east.

Object	Page
Pi Andromedae	68
Andromeda Galaxy	69
NGC 752	70
Almach	71
Achird	72
Owl Cluster	73
Messier 103	74
NGC 663	75

Object	Page
Double Cluster	78
Messier 34	79
Mesarthim	76
Lambda Arietis	77
Pleiades	81
Crab Nebula	82
Messier 37	88
Messier 35	89

Object	Page
Castor	90
Meissa	84
Mintaka	83
Sigma Orionis	86
Orion Nebula	85
Gamma Leporis	87
Sirius	91
Beta Monocerotis	91

Chart 6

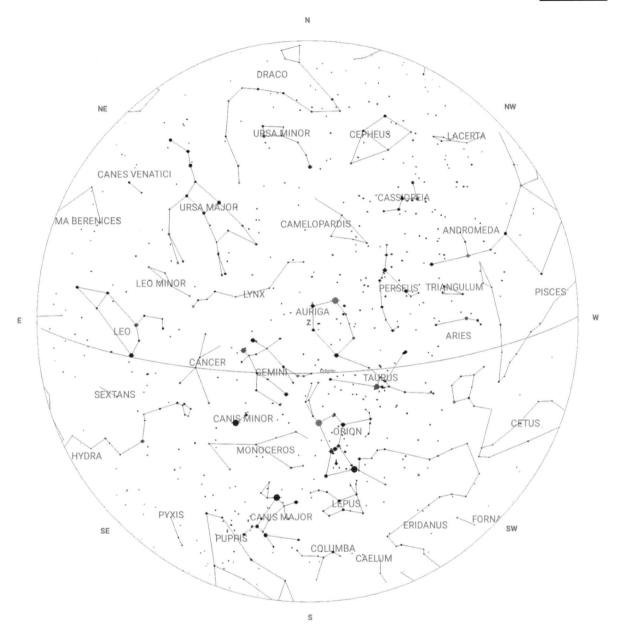

NB: Not all visible objects may be listed. Objects are listed based upon their visibility, from west to east.

Object	Page	Object	Page	Object	Page
Andromeda Galaxy	69	Messier 34	79	Meissa	84
NGC 752	70	Mesarthim	76	Mintaka	83
Almach	71	Lambda Arietis	77	Sigma Orionis	86
Achird	72	Pleiades	81	Orion Nebula	85
Owl Cluster	73	Crab Nebula	82	Gamma Leporis	87
Messier 103	74	Messier 37	88	Sirius	92
NGC 663	75	Messier 35	89	Messier 41	93
Double Cluster	78	Castor	90	Beta Monocerotis	91

Chart 7

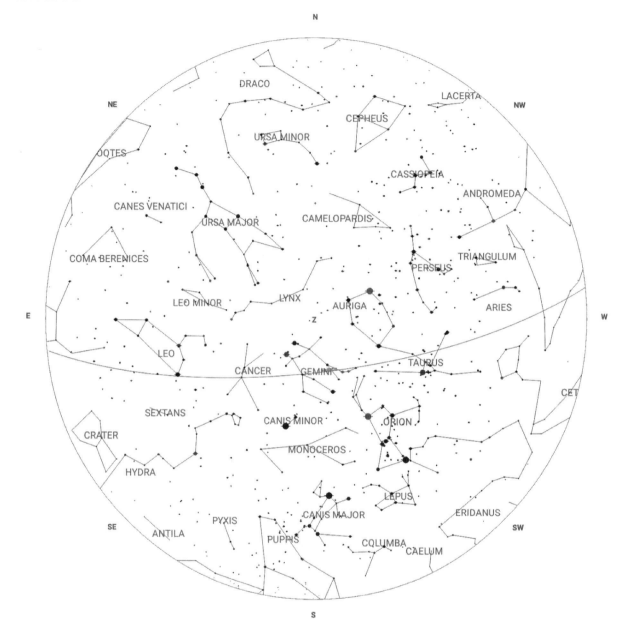

NB: Not all visible objects may be listed. Objects are listed based upon their visibility, from west to east.

Object	Page
Achird	72
Owl Cluster	73
Messier 103	74
NGC 663	75
Double Cluster	78
Messier 34	79
Pleiades	81
Crab Nebula	82

Object	Page
Messier 37	88
Messier 35	89
Castor	90
Meissa	84
Mintaka	83
Sigma Orionis	86
Orion Nebula	85
Gamma Leporis	87

Object	Page
Sirius	92
Messier 41	93
Messier 93	95
Beta Monocerotis	91
Procyon	94
Iota Cancri	97
Praesepe	96
Messier 67	98

Chart 8

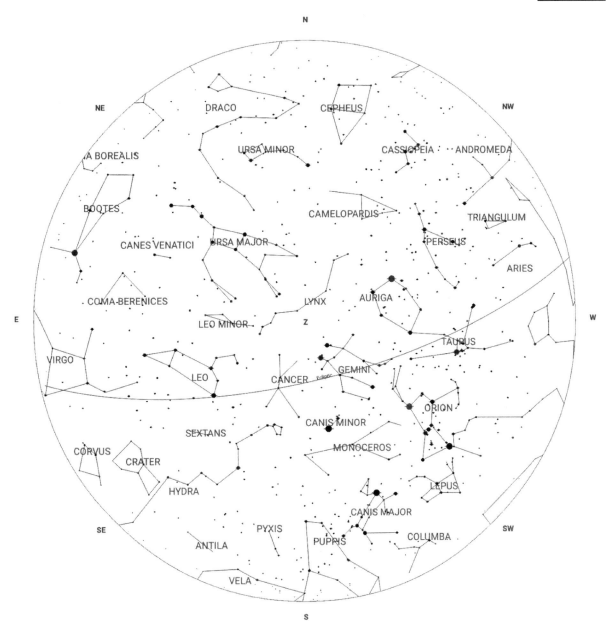

NB: Not all visible objects may be listed. Objects are listed based upon their visibility, from west to east.

Object	Page
Double Cluster	78
Messier 34	79
Pleiades	81
Crab Nebula	82
Messier 37	88
Messier 35	89
Castor	90
Meissa	84

Object	Page
Mintaka	83
Sigma Orionis	86
Orion Nebula	85
Gamma Leporis	87
Sirius	92
Messier 41	93
Messier 93	95
Beta Monocerotis	91

Object	Page
Procyon	94
Iota Cancri	97
Praesepe	96
Messier 67	98
Regulus	99
Algieba	101
Adhafera	100
Denebola	102

Chart 9

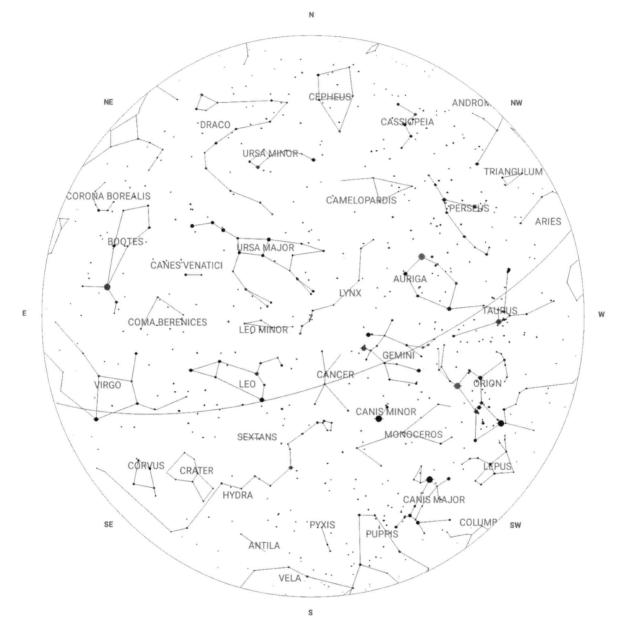

NB: Not all visible objects may be listed. Objects are listed based upon their visibility, from west to east.

Object	Page	Object	Page	Object	Page
Pleiades	81	Orion Nebula	85	Messier 67	98
Crab Nebula	82	Sirius	92	Regulus	99
Messier 37	88	Messier 41	93	Algieba	101
Messier 35	89	Messier 93	95	Adhafera	100
Castor	90	Beta Monocerotis	91	Denebola	102
Meissa	84	Procyon	94	Mizar & Alcor	105
Mintaka	83	Iota Cancri	97	Cor Caroli	104
Sigma Orionis	86	Praesepe	96	Arcturus	106

<u>Chart 10</u>

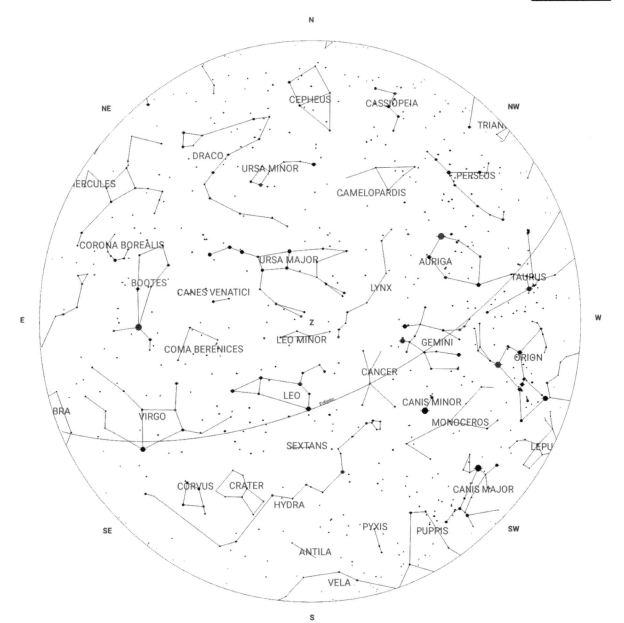

NB: Not all visible objects may be listed. Objects are listed based upon their visibility, from west to east.

Object	Page
Messier 37	88
Messier 35	89
Castor	90
Meissa	84
Sirius	92
Messier 41	93
Messier 93	95

Object	Page
Beta Monocerotis	91
Procyon	94
Iota Cancri	97
Praesepe	96
Messier 67	98
Regulus	99
Algieba	101

Object	Page
Adhafera	100
Denebola	102
Polaris	80
Mizar & Alcor	105
Cor Caroli	104
Arcturus	106
Delta Boötis	107

Chart 11

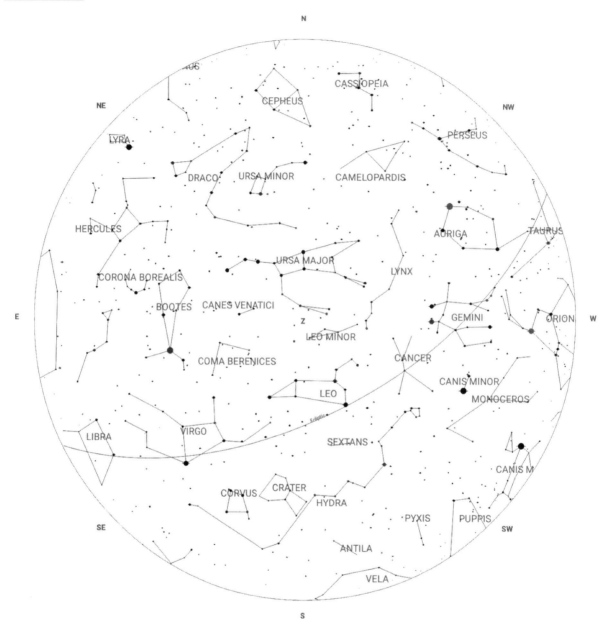

NB: Not all visible objects may be listed. Objects are listed based upon their visibility, from west to east.

Object	Page
Messier 37	88
Messier 35	89
Castor	90
Beta Monocerotis	91
Procyon	94
Iota Cancri	97
Praesepe	96

Object	Page
Messier 67	98
Regulus	99
Algieba	101
Adhafera	100
Denebola	102
Algorab	103

Object	Page
Polaris	80
Kuma	117
Mizar & Alcor	105
Cor Caroli	104
Arcturus	106
Delta Boötis	107

Chart 12

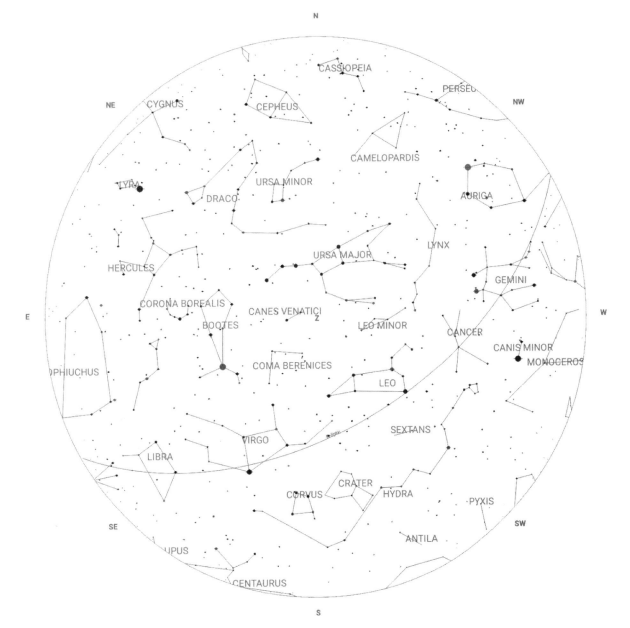

NB: Not all visible objects may be listed. Objects are listed based upon their visibility, from west to east.

Object	Page
Castor	90
Procyon	94
Iota Cancri	97
Praesepe	96
Messier 67	98
Regulus	99

Object	Page
Algieba	101
Adhafera	100
Denebola	102
Algorab	103
Polaris	80
Mizar & Alcor	105

Object	Page
Cor Caroli	104
Arcturus	106
Delta Boötis	107
Kuma	117
Keystone Cluster	115
Rho Herculis	116

Chart 13

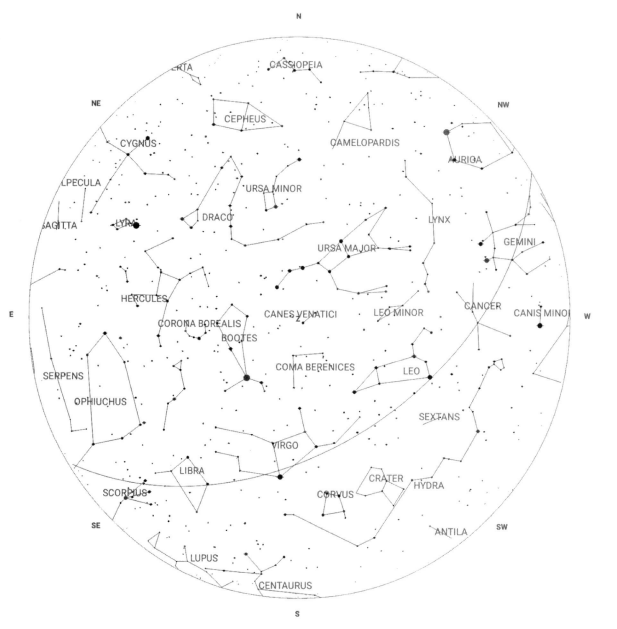

NB: Not all visible objects may be listed. Objects are listed based upon their visibility, from west to east.

Object	Page
Castor	90
Iota Cancri	97
Praesepe	96
Messier 67	98
Regulus	99
Algieba	101

Object	Page
Adhafera	100
Denebola	102
Algorab	103
Polaris	80
Mizar & Alcor	105
Cor Caroli	104

Object	Page
Arcturus	106
Delta Boötis	107
Keystone Cluster	115
Rho Herculis	116
Kuma	117
Zuben Elgenubi	108

<u>Chart 14</u>

NB: Not all visible objects may be listed. Objects are listed based upon their visibility, from west to east.

Object	Page
Iota Cancri	97
Praesepe	96
Messier 67	98
Regulus	99
Algieba	101
Adhafera	100
Denebola	102

Object	Page
Algorab	103
Polaris	80
Mizar & Alcor	105
Cor Caroli	104
Arcturus	106
Delta Boötis	107

Object	Page
Zuben Elgenubi	108
Keystone Cluster	115
Rho Herculis	116
Kuma	117
Double Double	119
Sheliak	120

Chart 15

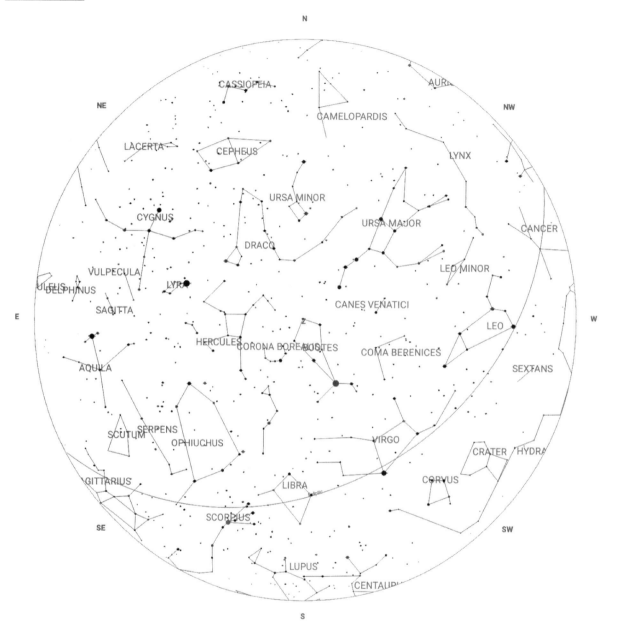

NB: Not all visible objects may be listed. Objects are listed based upon their visibility, from west to east.

Object	Page
Regulus	99
Algieba	101
Adhafera	100
Denebola	102
Algorab	103
Polaris	80
Mizar & Alcor	105

Object	Page
Cor Caroli	104
Arcturus	106
Delta Boötis	107
Zuben Elgenubi	108
Graffias	109
Jabbah	110
Keystone Cluster	115

Object	Page
Rho Herculis	116
Kuma	117
Double Double	119
Sheliak	120
Ring Nebula	121
Albireo	125
Messier 29	126

Chart 16

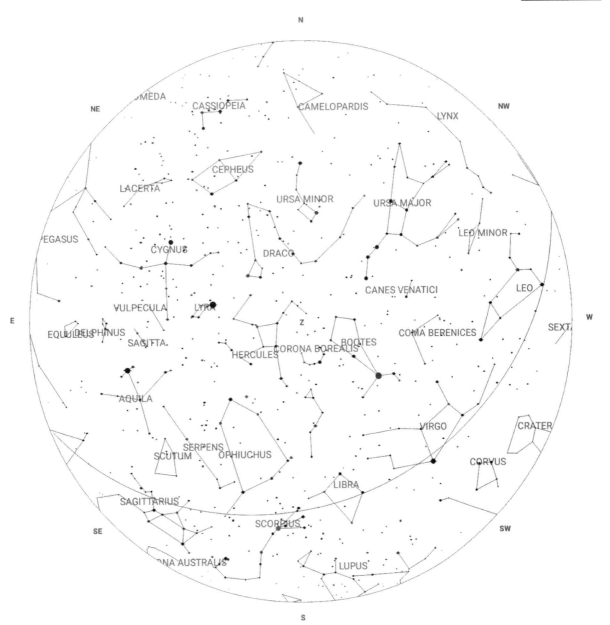

NB: Not all visible objects may be listed. Objects are listed based upon their visibility, from west to east.

Object	Page
Denebola	102
Polaris	80
Mizar & Alcor	105
Cor Caroli	104
Arcturus	106
Delta Boötis	107
Zuben Elgenubi	108
Graffias	109

Object	Page
Jabbah	110
Messier 80	111
Messier 4	112
Keystone Cluster	115
Rho Herculis	116
Kuma	117
Double Double	119
Sheliak	120

Object	Page
Ring Nebula	121
Albireo	125
Messier 29	126
Dumbbell Nebula	124
The Coathanger	123
Zeta Sagittae	127
Messier 71	128
Wild Duck Cluster	122

Chart 17

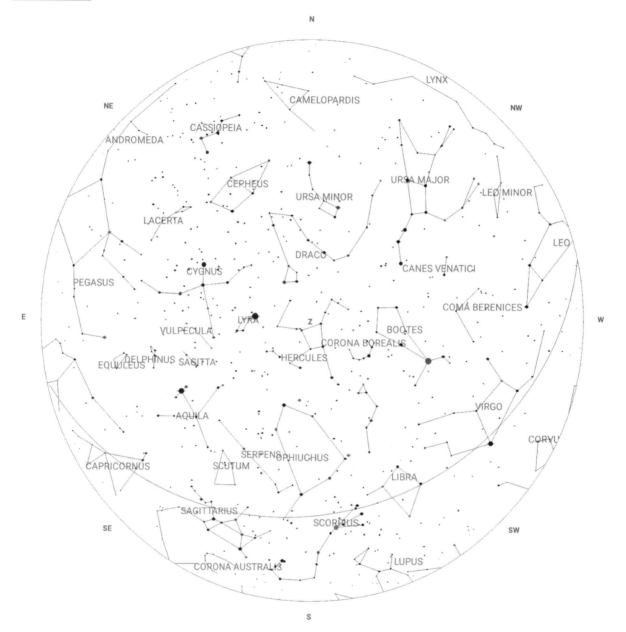

NB: Not all visible objects may be listed. Objects are listed based upon their visibility, from west to east.

Object	Page
Polaris	80
Mizar & Alcor	105
Cor Caroli	104
Arcturus	106
Delta Boötis	107
Zuben Elgenubi	108
Graffias	109
Jabbah	110

Object	Page
Messier 80	111
Messier 4	112
Keystone Cluster	115
Rho Herculis	116
Kuma	117
Double Double	119
Sheliak	120
Ring Nebula	121

Object	Page
Albireo	125
Messier 29	126
Dumbbell Nebula	124
The Coathanger	123
Zeta Sagittae	127
Messier 71	128
Gamma Delphini	130
Wild Duck Cluster	122

Chart 18

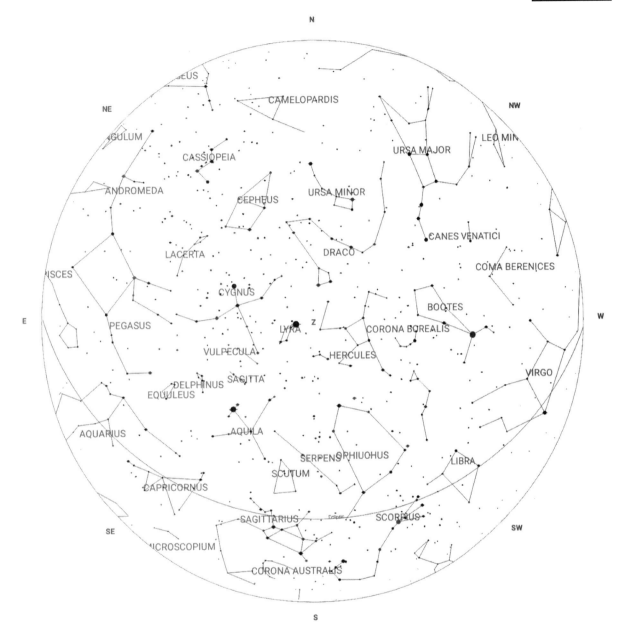

NB: Not all visible objects may be listed. Objects are listed based upon their visibility, from west to east.

Object	Page
Arcturus	106
Delta Boötis	107
Zuben Elgenubi	108
Graffias	109
Jabbah	110
Messier 80	111
Messier 4	112
Keystone Cluster	115

Object	Page
Rho Herculis	116
Kuma	117
Double Double	119
Sheliak	120
Ring Nebula	121
Albireo	125
Messier 29	126
Dumbbell Nebula	124

Object	Page
The Coathanger	123
Zeta Sagittae	127
Messier 71	128
Gamma Delphini	130
Wild Duck Cluster	122
Messier 22	118
Butterfly Cluster	113
Messier 7	114

Chart 19

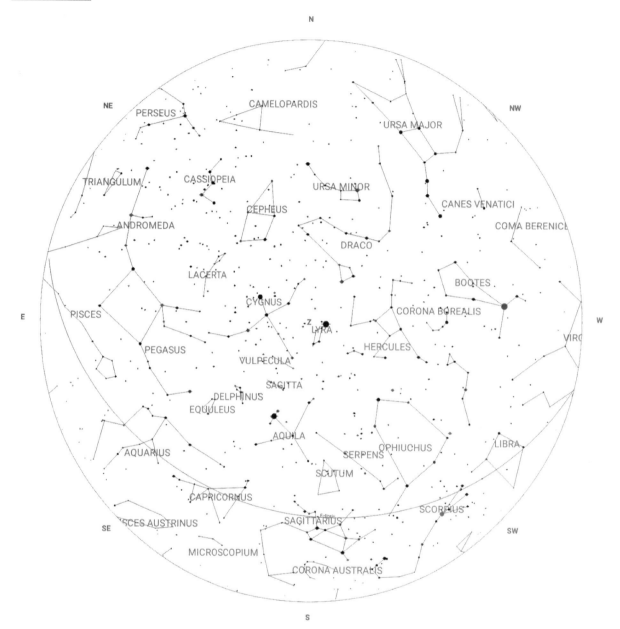

NB: Not all visible objects may be listed. Objects are listed based upon their visibility, from west to east.

Object	Page
Arcturus	106
Delta Boötis	107
Graffias	109
Jabbah	110
Keystone Cluster	115
Rho Herculis	116
Kuma	117
Double Double	119

Object	Page
Sheliak	120
Ring Nebula	121
Albireo	125
Messier 29	126
Dumbbell Nebula	124
The Coathanger	123
Zeta Sagittae	127
Messier 71	128

Object	Page
Gamma Delphini	130
Messier 15	131
Al Giedi	129
Wild Duck Cluster	122
Messier 22	118
Butterfly Cluster	113
Messier 7	114

Chart 20

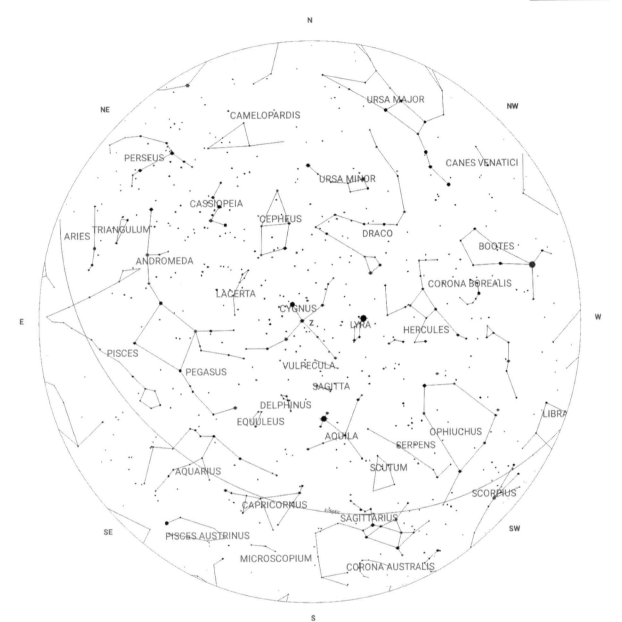

NB: Not all visible objects may be listed. Objects are listed based upon their visibility, from west to east.

Object	Page
Delta Boötis	107
Keystone Cluster	115
Rho Herculis	116
Kuma	117
Double Double	119
Sheliak	120
Ring Nebula	121
Albireo	125

Object	Page
Messier 29	126
Dumbbell Nebula	124
The Coathanger	123
Zeta Sagittae	127
Messier 71	128
Gamma Delphini	130
Messier 15	131
Al Giedi	129

Object	Page
Wild Duck Cluster	122
Messier 22	118
Pi Andromedae	68
Achird	72
Owl Cluster	73
Messier 103	74
NGC 663	75

Chart 21

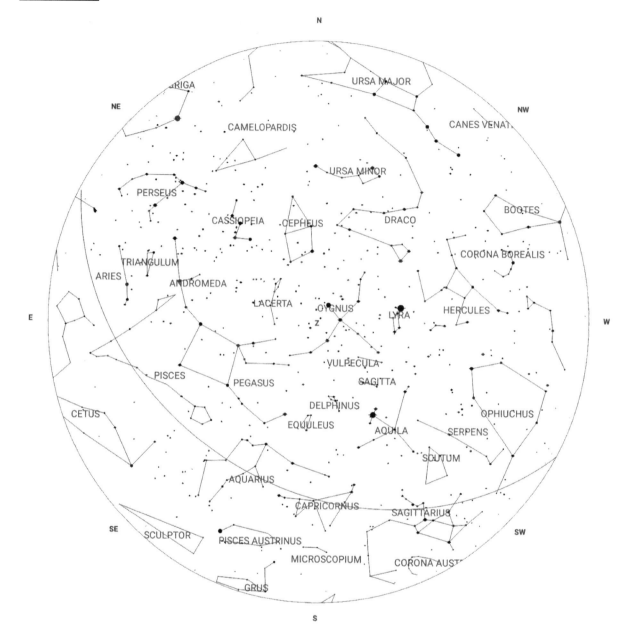

NB: Not all visible objects may be listed. Objects are listed based upon their visibility, from west to east.

Object	Page
Keystone Cluster	115
Rho Herculis	116
Kuma	117
Double Double	119
Sheliak	120
Ring Nebula	121
Albireo	125
Messier 29	126

Object	Page
Dumbbell Nebula	124
The Coathanger	123
Zeta Sagittae	127
Messier 71	128
Gamma Delphini	130
Messier 15	131
Al Giedi	129
Wild Duck Cluster	122

Object	Page
Pi Andromedae	68
Andromeda Galaxy	69
Almach	71
Achird	72
Owl Cluster	73
Messier 103	74
NGC 663	75

Chart 22

NB: Not all visible objects may be listed. Objects are listed based upon their visibility, from west to east.

Object	Page
Keystone Cluster	115
Rho Herculis	116
Kuma	117
Double Double	119
Sheliak	120
Ring Nebula	121
Albireo	125
Messier 29	126

Object	Page
Dumbbell Nebula	124
The Coathanger	123
Zeta Sagittae	127
Messier 71	128
Gamma Delphini	130
Messier 15	131
Al Giedi	129
Wild Duck Cluster	122

Object	Page
Pi Andromedae	68
Andromeda Galaxy	69
Almach	71
Achird	72
Owl Cluster	73
Messier 103	74
NGC 663	75
Double Cluster	78

Chart 23

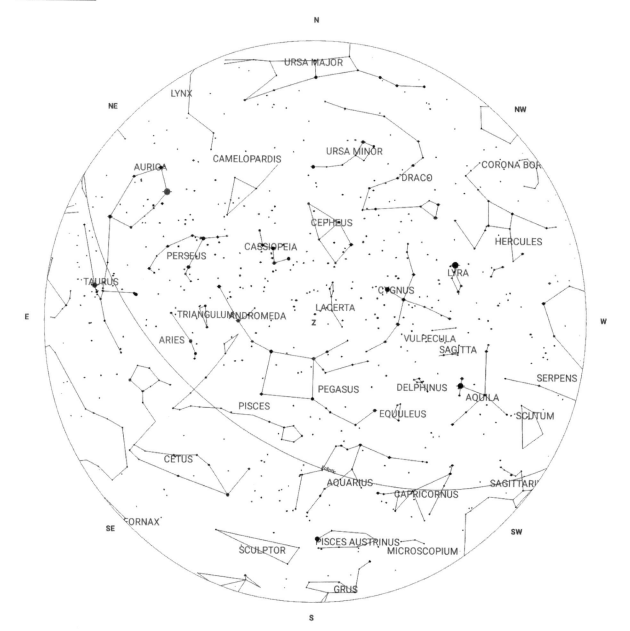

NB: Not all visible objects may be listed. Objects are listed based upon their visibility, from west to east.

Object	Page
Kuma	117
Double Double	119
Sheliak	120
Ring Nebula	121
Albireo	125
Messier 29	126
Dumbbell Nebula	124
The Coathanger	123

Object	Page
Zeta Sagittae	127
Messier 71	128
Gamma Delphini	130
Messier 15	131
Pi Andromedae	68
Andromeda Galaxy	69
Almach	71
Achird	72

Object	Page
Owl Cluster	73
Messier 103	74
NGC 663	75
Double Cluster	78
Messier 34	79
NGC 752	70
Lambda Arietis	77
Mesarthim	76

Chart 24

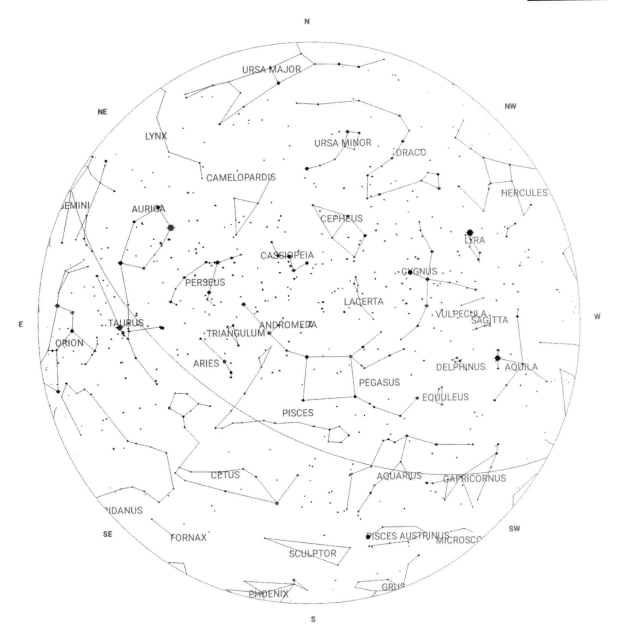

NB: Not all visible objects may be listed. Objects are listed based upon their visibility, from west to east.

Object	Page
Double Double	119
Sheliak	120
Ring Nebula	121
Albireo	125
Messier 29	126
Dumbbell Nebula	124
The Coathanger	123
Zeta Sagittae	127

Object	Page
Messier 71	128
Gamma Delphini	130
Messier 15	131
Pi Andromedae	68
Andromeda Galaxy	69
Almach	71
Achird	72
Owl Cluster	73

Object	Page
Messier 103	74
NGC 663	75
Double Cluster	78
Messier 34	79
NGC 752	70
Lambda Arietis	77
Mesarthim	76
Pleiades	81

The Deep Sky Objects

Pi Andromedae

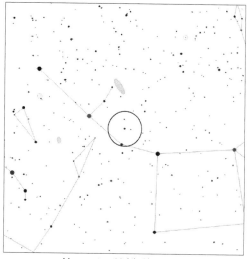

Designation(s):	Pi Andromedae
Constellation:	Andromeda
R.A.:	00h 36m 53s
Declination:	+33° 43' 10"
Object Type:	Multiple Star
Location:	★ ★
Rating:	★ ★
Best Seen:	Autumn and Winter

Map courtesy *Mobile Observatory*

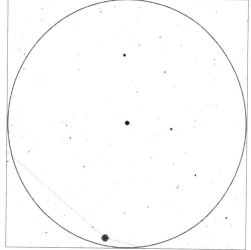

Finderscope view courtesy *Mobile Observatory*

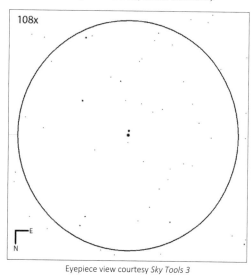

Eyepiece view courtesy *Sky Tools 3*

Pi Andromedae is not one of the brightest stars in the constellation but is still quite easily found from suburban or dark sky locations.

Andromeda is always depicted as being a curved line of stars extending east from the square of Pegasus. However, if you look carefully, you'll see a second curved line of fainter stars of which Pi is the first. (The second, Mu Andromedae, is fairly close to the Andromeda Galaxy.)

Binoculars won't split the star but a small telescope at low power should easily do the trick. At 26x I saw a bright white primary with a very faint bluish secondary. From the city I noted that the secondary appeared to have a coppery color, but I suspect that's more likely due to the city air.

There is a third, much fainter and wider companion but I've been unable to see it from either the suburbs or from the city.

The components we see with our telescopes are not members of a true multiple star system but are simply a chance alignment of stars at differing distances from us. However, the primary is actually a very close pair of nearly identical stars that each shines with the light of a thousand Suns. The pair orbit one another once every 144 days and are approximately 600 light years away.

The Andromeda Galaxy

Designation(s):	Messier 31
Constellation:	Andromeda
R.A.:	00h 42m 44s
Declination:	+41° 16' 09"
Object Type:	Spiral Galaxy
Location:	★ ★
Rating:	★ ★
Best Seen:	Autumn and Winter

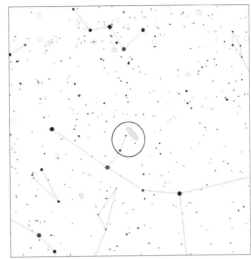

Map courtesy *Mobile Observatory*

Known since antiquity, the Andromeda Galaxy is one of the most fascinating objects you can see – but if you own a small telescope, it can also be one of the most disappointing.

The reason is simple. Although it's visible to the unaided eye under dark skies, its light is spread over a large area and, consequently, it usually doesn't appear that bright (or detailed) through a small telescope.

So what's the fascination? Well, for one thing, you're looking at a whole other galaxy.

Our own Milky Way galaxy has about 200 billion stars and the Andromeda Galaxy is even bigger. And at a distance of 2 ½ million light years, it's the most distant object you can easily see with just your eyes. It's also worth remembering that, as you stare at it through the eyepiece, the light that's now hitting your eyes first started out millions of years ago.

What does this mean? You are, in fact, looking at the galaxy as it once was. You're literally looking into the past.

Under suburban skies at 35x, it appears as a moderately bright, misty oval but in the city it was hard to see at all. In fact, I had to use averted vision to even see its shape. Otherwise, only the core is visible. Look for it, enjoy it, but don't get your hopes up too much!

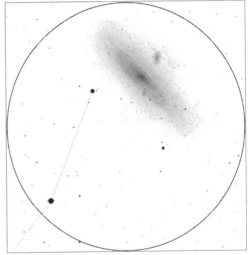

Finderscope view courtesy *Mobile Observatory*

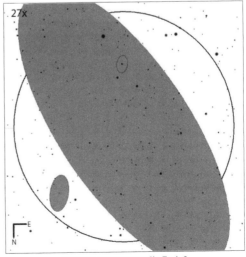

Eyepiece view courtesy *Sky Tools 3*

NGC 752

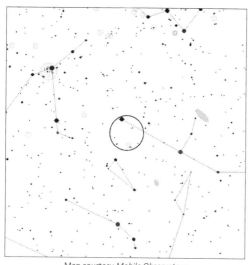

Map courtesy *Mobile Observatory*

Designation(s):	NGC 752
Constellation:	Andromeda
R.A.:	01h 57m 55s
Declination:	+37° 51' 57"
Object Type:	Open Cluster
Location:	★ ★
Rating:	★ ★
Best Seen:	Autumn and Winter

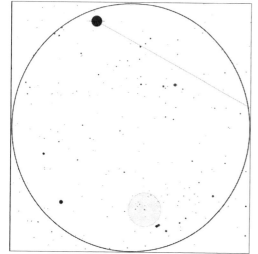

Finderscope view courtesy *Mobile Observatory*

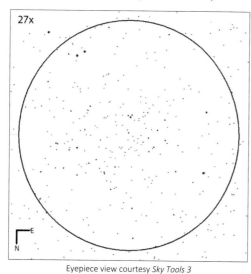

Eyepiece view courtesy *Sky Tools 3*

NGC 752 is one of the larger open clusters and its light is consequently scattered across a relatively large area of the sky.

That being said, it should be visible in binoculars as a faint misty patch and using averted vision may help to resolve some individual stars.

This is one of those clusters that looks good under both suburban and city-based skies. You should be able to locate it with Almach on one edge of the finderscope field of view and the cluster on the opposite edge.

At 35x, it appears as a large, sparse cluster and denser toward the core. The stars are predominantly blue-white but you'll also notice some scattered orange stars throughout.

The double star 56 Andromedae lies on the edge and makes a convenient marker for the cluster. When viewed at this magnification, the entire cluster fits into the field of view with an orange star lying just outside the main group.

On occasion, when observing the cluster as it sets in the west, it takes on the appearance of Boötes with this star taking the place of Arcturus.

Designation(s):	Gamma Andromedae
Constellation:	Andromeda
R.A.:	02h 03m 54s
Declination:	+42° 19' 47"
Object Type:	Multiple Star
Location:	★ ★ ★
Rating:	★ ★ ★
Best Seen:	Autumn and Winter

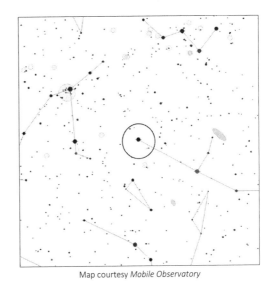

Map courtesy *Mobile Observatory*

Almach, the third brightest star in Andromeda, is a true multiple star system located some 350 light years away. Although only two stars can easily be seen with a small telescope, there are, in fact, four stars present with the fainter, blue component being its own triple star system.

Definitely a favorite for the autumn, this double star can be split with low power and is reminiscent of Albireo in Cygnus.

But whereas Albireo is easily resolved at low power, you'll probably need at least 50x to properly enjoy the view. I've barely split it at 26x from the city, but this is one double where greater magnification definitely comes in handy. I've found that a magnification of about 65x or 70x works quite nicely.

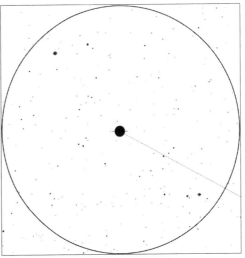

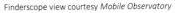

Finderscope view courtesy *Mobile Observatory*

The primary is a pale yellow-white gold and about four times brighter than the pale blue secondary. On a few occasions, I've noted that the secondary will flash purple and violet, but these colors may only be apparent at higher magnification (around 90x or 100x.)

One last thing – I've noticed that Almach is one of those pairs where the colors were more apparent at higher magnifications. What do you see?

Almach was first seen as a double in 1778 by the German physicist Johann Tobias Mayer.

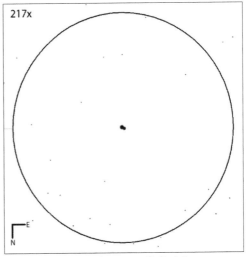

Eyepiece view courtesy *Sky Tools 3*

Achird

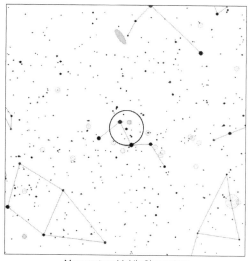

Map courtesy *Mobile Observatory*

Designation(s):	Eta Cassiopeiae
Constellation:	Cassiopeia
R.A.:	00h 49m 06s
Declination:	+57° 48' 55"
Object Type:	Multiple Star
Location:	★ ★ ★
Rating:	★ ★ ★
Best Seen:	Autumn and Winter

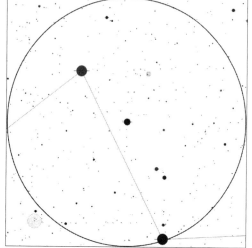

Finderscope view courtesy *Mobile Observatory*

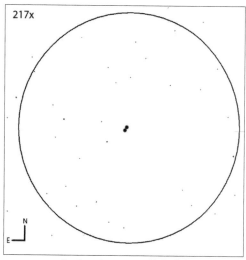

Eyepiece view courtesy *Sky Tools 3*

Achird is a relatively close multiple star; many sources will mention a gold primary with a reddish-purple companion, but I've noted other colors and a third, bluish star nearby.

Although it doesn't require much power to reveal the two main components, I've been unable to split it at a magnification of 26x but upping it a little to 35x or 40x should do the trick.

The first, and brightest, component appears a brilliant white to me (rather than the gold noted by others) but you'll need to focus carefully to reveal the much fainter companion.

At this magnification, it's difficult for me to notice any particular color but increasing the magnification again, to about 50x, helps to reveal the companion's coppery complexion.

While you have it at about 50x, try looking for the third component, which should appear as a tiny blue star on the opposite side of the white primary.

If not, increase the magnification once again. I've been able to spot it at 65x but 87x seemed to provide a particularly nice view with all three stars visible.

The Owl Cluster

Designation(s):	NGC 457
Constellation:	Cassiopeia
R.A.:	01h 19m 33s
Declination:	+58° 17' 27"
Object Type:	Open Cluster
Location:	★
Rating:	★ ★ ★
Best Seen:	Autumn and Winter

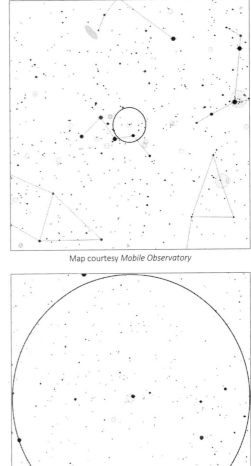

Map courtesy *Mobile Observatory*

A definite favorite, this cluster has a very distinctive shape that has given rise to a number of imaginative names over the years. Many see the stars forming the shape of an owl, while others see a kite. More recently, some have come to call it the E.T. Cluster.

Although the cluster can be seen with binoculars it doesn't truly shine until you turn your telescope toward it. You won't need a lot of magnification either – it can be seen with 26x and 35x provides a nice view, but somewhere between 50x and 90x is probably best. Once you get to about 100x, you'll have difficulty fitting the entire cluster into the field of view.

Finderscope view courtesy *Mobile Observatory*

The first time you see it, you're sure to be delighted as your eyes take in the sight and your imagination takes over. With the double star Phi Cassiopeiae marking the eyes, it's easy to see the cluster as either an owl with powerful wings or an alien with long, outstretched arms staring back at you.

The two stars of Phi Cassiopeiae are both white, with one being about 1½ times brighter than the other. It's a sparsely scattered cluster with the densest portion being around the chest area of the alien.

If the thought of an alien spooks you, look again. No owl? How about a goose or a swan in flight with Phi marking the tail?

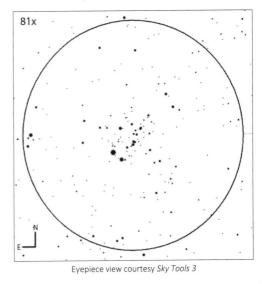

Eyepiece view courtesy *Sky Tools 3*

Messier 103

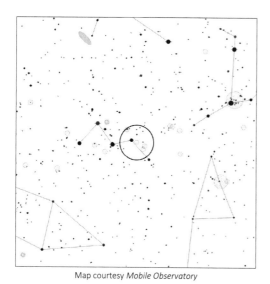

Map courtesy *Mobile Observatory*

Designation(s):	Messier 103
Constellation:	Cassiopeia
R.A.:	01h 33m 22s
Declination:	+60° 39' 29"
Object Type:	Open Cluster
Location:	★
Rating:	★
Best Seen:	Autumn and Winter

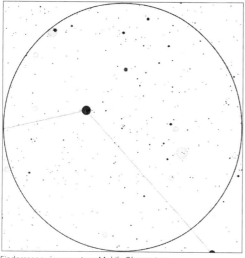

Finderscope view courtesy *Mobile Observatory*

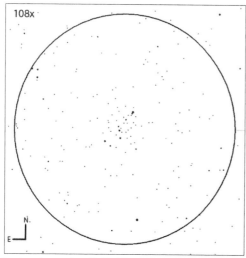

Eyepiece view courtesy *Sky Tools 3*

Messier 103 is a small open cluster that's easily found but may be a little disappointing under city and suburban skies. It's barely visible as a tiny, compact elongated patch in binoculars but you may still be able to see the three stars that make up the slightly curved line across the center.

The telescopic view doesn't improve much from the city. I was only able to see about four or five stars at 26x – the central three for sure – with the center star appearing reddish.

The cluster fares better under suburban skies but it's best to observe it with low expectations. At 35x I could easily see the three bright stars across the center, like a mini Orion's belt. The middle star appeared to be golden with a blue companion close by while the other stars appeared to be white. I also noticed a triangle of stars to the north-east.

Upping the magnification to 91x definitely helped. The cluster still fitted within the field of view and I was able to see a lot more stars with averted vision.

There's one simple reason why the cluster appears so small: at a distance of about 9,000 light years, Messier 103 is one of the most distant open clusters known. There are thought to be over 150 member stars with the cluster being about 25 million years old.

Designation(s):	NGC 663
Constellation:	Cassiopeia
R.A.:	01h 46m 16s
Declination:	+61° 13′ 06″
Object Type:	Open Cluster
Location:	★ ★
Rating:	★ ★ ★
Best Seen:	Autumn and Winter

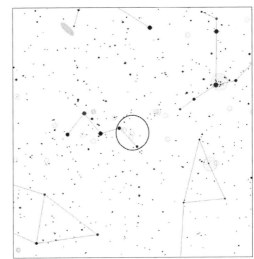

Map courtesy *Mobile Observatory*

NGC 663 is a little gem of a cluster. Like the Owl Cluster, it's easily found and is quite large and attractive, which leads you to wonder why Messier didn't include it in his famous list.

This cluster can be easily seen with binoculars as a large, faint, hazy circular patch about twice the size of Messier 103. It looks like a sparsely scattered globular cluster and you may be able to see individual stars with averted vision.

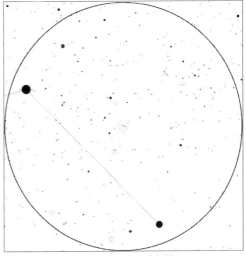

Finderscope view courtesy *Mobile Observatory*

However, it's best observed with a telescope at low to medium power, somewhere between 50x and 100x.

At 35x it appeared to be a sparsely scattered collection of stars that formed a diamond shape. The vast majority of stars were blue-white, but there were two pairings of stars – one just off center and the other near the edge – that appeared to be orange.

The two pairs of stars are actually a little spooky and make me think of two pairs of eyes. With the cluster high in the sky at Halloween, maybe they're ghosts. Or if NGC 457 is the E.T. Cluster, maybe E.T. is just the largest and friendliest of the aliens and NGC 663 are a smaller pair following close behind.

The cluster lies about 7,000 light years away and contains some 400 stars.

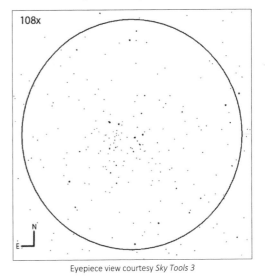

Eyepiece view courtesy *Sky Tools 3*

Mesarthim

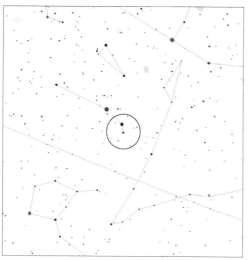

Map courtesy *Mobile Observatory*

Designation(s):	Gamma Arietis
Constellation:	Aries
R.A.:	01h 53m 32s
Declination:	+19° 17' 38"
Object Type:	Multiple Star
Location:	★ ★ ★
Rating:	★ ★ ★
Best Seen:	Autumn and Winter

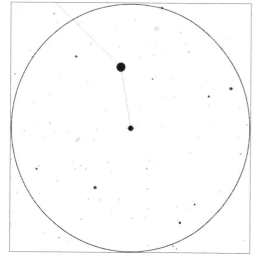

Finderscope view courtesy *Mobile Observatory*

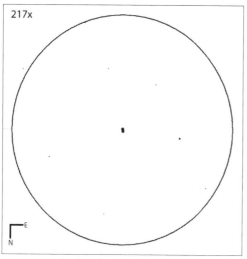

Eyepiece view courtesy *Sky Tools 3*

Mesarthim is a popular and relatively easy double star, known to many amateur astronomers across the world.

It's easily found and a good target if you're new to the hobby, especially as it can be used as a good benchmark for your equipment and the seeing conditions.

For example, you should be able to split the star at a fairly low magnification. From the city, I can barely split it at 26x with my 130mm scope but at 35x it's more apparent. I usually had no trouble splitting the star at that magnification from the suburbs with my 4 ½ inch XT.

How cleanly the star is split can depend upon the atmosphere; if the air isn't steady, it might appear to be barely split but otherwise it should be pretty clean.

You'll want to increase the magnification to get the best view; somewhere around 100x is probably best but even at lower magnification you should be able to see a third, unrelated fainter star nearby. What color is that star? To me it appears blue.

There's no mistaking the color of the Mesarthim stars themselves as both components are pure white and of equal brightness. Discovered by Robert Hooke in 1664, the two stars orbit one another every 5,000 years and lie about 165 light years away from us.

Lambda Arietis

Designation(s):	Lambda Arietis
Constellation:	Aries
R.A.:	01h 57m 56s
Declination:	+23° 35' 46"
Object Type:	Multiple Star
Location:	★ ★ ★
Rating:	★ ★ ★
Best Seen:	Autumn and Winter

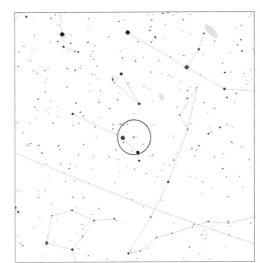

Map courtesy *Mobile Observatory*

Lambda Arietis is slightly off the beaten track but should still be quite easily found. It lies between two brighter stars – Hamal to the east and Sheratan to the west. Lambda appears slightly closer to Hamal in the map and finderscope views depicted here.

Some observers have reported this as being split with binoculars, but I've had no luck with my 10x50's.

However, a low magnification of 26x should easily reveal a brilliant white star that's about twice as bright as the bluish secondary. (On one occasion I noted hints of violet in the companion but the color seemed to disappear with a higher power eyepiece.)

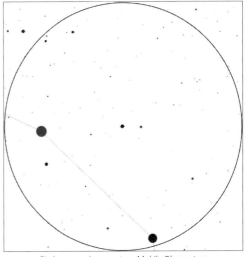

Finderscope view courtesy *Mobile Observatory*

A magnification of about 65x will provide a nice view. Look out for a third, slightly fainter companion nearby that forms an elongated triangle with the main pair.

Lambda is thought to be a true double star system as both stars appear to be moving together through space. In reality, both stars are actually yellow but the primary is hotter and its color is closer to yellow-white.

The reason the secondary appears blue is because your eyes are sensitive to contrasts in light and you see an exaggerated contrast effect as a result.

The stars lie at a distance of about 129 light years and take more than 33,000 years to orbit one another.

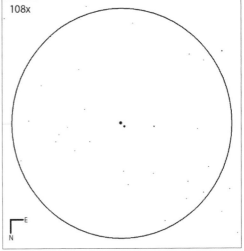

Eyepiece view courtesy *Sky Tools 3*

The Double Cluster

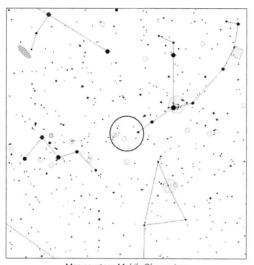

Map courtesy *Mobile Observatory*

Designation(s):	NGC 869 & NGC 884
Constellation:	Perseus
R.A.:	02h 22m 29s
Declination:	+57° 09′ 00″
Object Type:	Open Cluster
Location:	★ ★
Rating:	★ ★ ★
Best Seen:	Autumn and Winter

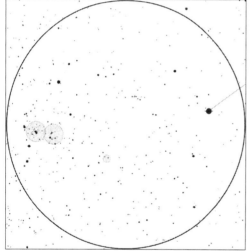

Finderscope view courtesy *Mobile Observatory*

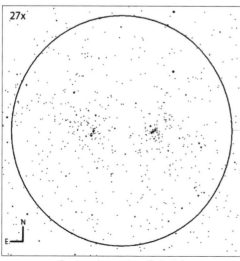

Eyepiece view courtesy *Sky Tools 3*

Known since antiquity, the Double Cluster is one of the highlights of the night sky and yet, somehow, like NGC 663, Messier never included it in his catalog. They are, quite simply, unmissable in every sense.

Depending on the quality of your sky, you may be able to glimpse this stellar pair with just your eyes. Look about midway between the top of Perseus, the Hero and Cassiopeia, the Queen and you might see a faint, elongated patch of hazy light.

If you don't have any luck, try scanning the area with binoculars or your finderscope as they should be easily seen in both. To me, in my small 8x30 binoculars, the clusters seemed to take on the form of a butterfly or bowtie.

These clusters appear fairly large in the sky and are consequently best viewed at low power but I've been able to fit them both into the same field of view up to about 70x or so.

NGC 884, the eastern cluster (to the left in the eyepiece depiction) is larger and more sparsely scattered than its neighbor. NGC 869 has a dense core with two bright blue-white stars, just off-center. Take your time with these clusters and allow your eyes to soak in the view. Most of the stars will be young and blue-white in color but you'll see a few older orange giants too.

Designation(s):	Messier 34
Constellation:	Perseus
R.A.:	02h 42m 07s
Declination:	+42° 44' 46"
Object Type:	Open Cluster
Location:	★ ★
Rating:	★ ★
Best Seen:	Autumn and Winter

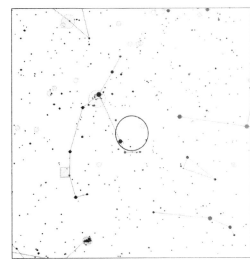

Map courtesy *Mobile Observatory*

Messier 34 is like a cousin to the Double Cluster and is often overlooked in favor of its flashier neighbors.

It can be found by pointing your finderscope or binoculars midway between Almach in Andromeda and Algol in Perseus. By placing Algol on the edge of the finderscope view, you should be able to see M34 on the opposite edge.

This is a fairly large, sparse cluster that reminds me of the Praesepe (M44) and is best viewed at low power. I found a magnification of about 50x to be best.

At 35x I noted that the cluster appeared to be comprised almost entirely of blue-white stars but with a few that show hints of amber. In particular, I've noticed a golden star on its western edge.

There are also a number of apparent double stars scattered throughout with a pair of equal brightness near a triangle at the center.

Do you notice any shapes? I saw an elongated X at 26x and I've also noted a pattern that's reminiscent of the Owl Cluster at the heart of the cluster.

The cluster lies about 1,500 light years away and is thought to contain close to 500 stars.

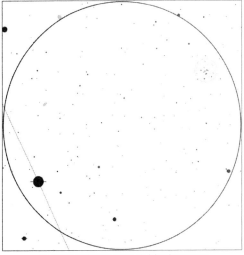

Finderscope view courtesy *Mobile Observatory*

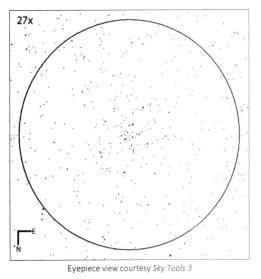

Eyepiece view courtesy *Sky Tools 3*

Polaris

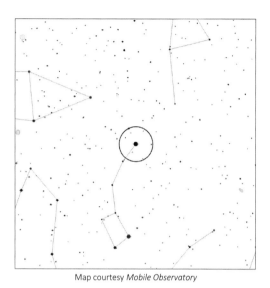

Map courtesy *Mobile Observatory*

Designation(s):	Alpha Ursa Minoris
Constellation:	Ursa Minor
R.A.:	02h 31m 49s
Declination:	+89° 15' 51"
Object Type:	Multiple Star
Location:	★ ★ ★
Rating:	★
Best Seen:	All Year Round

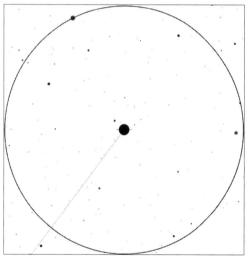

Finderscope view courtesy *Mobile Observatory*

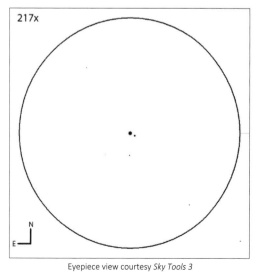

Eyepiece view courtesy *Sky Tools 3*

It's a common misconception among non-astronomers that Polaris, the pole star, is actually the brightest star in the sky. This has always baffled me as I can't imagine how this belief may have come about.

It's actually just a star of average brightness but it's very easily found and, since it's almost exactly at the north celestial pole, it never moves in the sky. (At least, not to the unaided eye.) It therefore has the convenience of being visible on every clear night of the year.

To be honest, it's not a very exciting multiple star but it's definitely one that should be on every newbie's bucket list. I haven't been able to split it at 26x and you may have to almost triple that power to about 75x to see the companion.

You should see a white primary with a very faint bluish secondary. It's an easy split at 130x but the secondary was difficult to see from the city and I needed to use averted vision to make it more apparent.

Once it's been split and you know where to look for the companion, gradually lower the magnification and see how low you can go before it disappears. Under suburban skies I've definitely seen it at 54x and have suspected the secondary at 35x.

The Pleiades

Designation(s):	Messier 45
Constellation:	Taurus
R.A.:	03h 47m 58s
Declination:	+24° 09' 50"
Object Type:	Open Cluster
Location:	★ ★ ★
Rating:	★ ★ ★
Best Seen:	Autumn and Winter

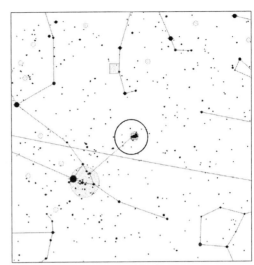

Map courtesy *Mobile Observatory*

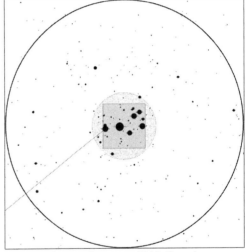

Finderscope view courtesy *Mobile Observatory*

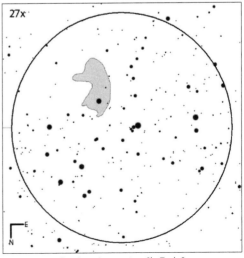

Eyepiece view courtesy *Sky Tools 3*

Astronomically speaking, winter is a great time for observers in the northern hemisphere. Not only can we admire the beautiful Orion Nebula, but we're also treated to the Pleiades, a stunning open star cluster in the constellation of Taurus the Bull.

It's a famous cluster that's been known by numerous names across the world for many, many years. In Japan, it's known as the Subaru (hence, the car manufacturer) while elsewhere it's known as the Seven Sisters. And this is something of a mystery because although the cluster is easily seen with just your eyes, most observers can only count six stars at most. So what happened to the seventh?

It's a large cluster (it appears twice the size of the full Moon in the sky) and, consequently, some say it's best observed with binoculars where it can be appreciated against the background sky.

However, if you turn a small telescope toward it with a low magnification eyepiece, you won't be disappointed. The entire cluster should fit into the field of view at about 35x with hundreds of blue-white stars scattered across the scene. Increasing the magnification might actually ruin the effect.

Lastly, if you get the opportunity, observe this cluster well away from any light pollution. You won't regret it.

The Crab Nebula

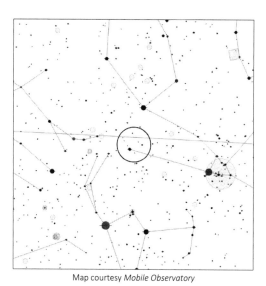

Map courtesy *Mobile Observatory*

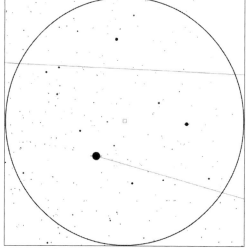

Finderscope view courtesy *Mobile Observatory*

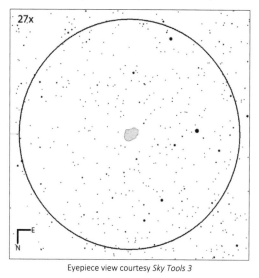

Eyepiece view courtesy *Sky Tools 3*

Designation(s):	Messier 1
Constellation:	Taurus
R.A.:	05h 34m 32s
Declination:	+22° 00' 52"
Object Type:	Supernova Remnant
Location:	★
Rating:	★
Best Seen:	Autumn and Winter

The Crab Nebula is thought to be the remains of a star that exploded almost a thousand years ago. On July 4th, 1054 Chinese astronomers recorded seeing a "guest star" in the sky. This supernova was so bright it was visible during daylight hours for about three weeks.

Nearly seven hundred years later, in 1731, the nebula was discovered by the astronomer John Bevis in the same area of the sky, but it wasn't linked to the supernova until the early 20th century.

To be honest, it can be disappointing when viewed in a small telescope, but it's worth remembering what you're looking at. There aren't too many supernovae remnants around and the Crab is the only example that can be readily observed by amateurs with inexpensive equipment.

Some can spot the nebula with binoculars but I've not had such luck and I suspect you'll need clear dark skies to be successful. However, it should be visible under suburban skies with a small telescope.

It's barely seen at about 35x but it's not an easy thing to see. Using averted vision, it appears as a very faint oval patch with little or no brightening near the center. I once described it as "a coal stain on a black carpet." Increasing the magnification may help, but if you live in the city, you'll probably be out of luck as I've never seen it from Los Angeles.

Designation(s):	Delta Orionis
Constellation:	Orion
R.A.:	05h 32m 00s
Declination:	-00° 17' 57"
Object Type:	Multiple Star
Location:	★ ★ ★
Rating:	★ ★
Best Seen:	Winter

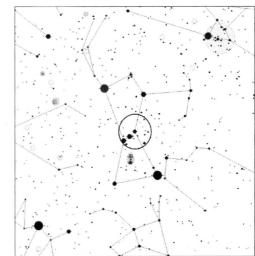

Map courtesy *Mobile Observatory*

Mintaka is a bright star, easily identified as being the most westerly star in Orion's belt. The belt has been a famous astronomical fixture for thousands of years with the three stars featuring prominently in cultures around the world.

But before you point your scope in its direction, scan the area with binoculars or a small telescope at low power. What can you see? The three belt stars are actually members of a very large open star cluster, known as the Orion OB1 Association and the area is studded with stars.

Mintaka itself is a fairly wide double, but unfortunately not so wide that you can split it with binoculars. (Or at least I've been unable to.)

Having said that, you won't need a lot of power to split it as about 26x should do the trick. At this magnification the two components are easily seen, with the white primary appearing about three or four times brighter than the sky blue secondary.

It looks better at higher magnification with about 50x providing a good view. Something I've noticed is that the higher the power, the stronger the colors appeared to be. Unfortunately, the pair may be less visually appealing as the gap between them appears to increase.

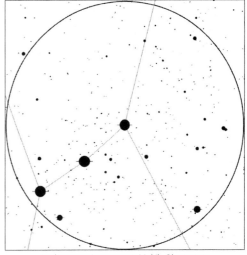

Finderscope view courtesy *Mobile Observatory*

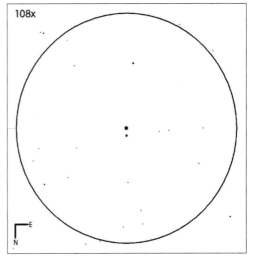

108x

Eyepiece view courtesy *Sky Tools 3*

Meissa

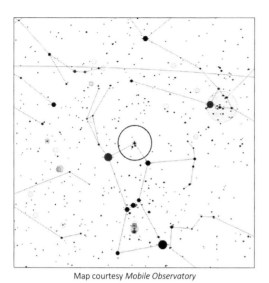

Map courtesy *Mobile Observatory*

Designation(s):	Lambda Orionis
Constellation:	Orion
R.A.:	05h 35m 08s
Declination:	+09° 56′ 03″
Object Type:	Multiple Star
Location:	★ ★ ★
Rating:	★ ★ ★
Best Seen:	Winter

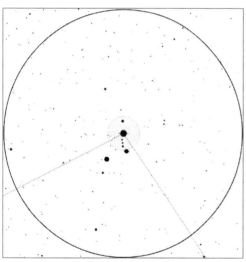

Finderscope view courtesy *Mobile Observatory*

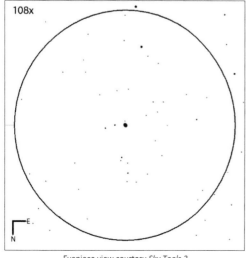

Eyepiece view courtesy *Sky Tools 3*

Meissa is the brightest member of a small cluster of stars that mark Orion's head. Through binoculars or a finderscope you should see a triangular asterism with a line of three faint stars close to Meissa itself.

At 26x, you'll see Meissa as a white star with a much fainter secondary nearby. If you look carefully with averted vision, you might be able to catch a third star between the two.

Up the power to about 35x and the third star should be easily seen, helping to form another line of three stars, equally spaced, with Meissa being at the end of the line and much brighter than the other two.

But that's not all. Increase the magnification again to about 70x and you might just be able to split Meissa itself. Stare for a few seconds and wait for the air to steady and a violet-blue companion will appear to be touching the star.

Meissa is one of those stars where a higher magnification will definitely benefit you. You'll need about 100x to cleanly split the pair (you might even see another, faint blue star nearby at this magnification) but the higher you go, the better the view.

Use your Barlow!

Designation(s):	Messier 42
Constellation:	Orion
R.A.:	05h 35m 17s
Declination:	-05° 23' 28"
Object Type:	Nebula
Location:	★ ★ ★
Rating:	★ ★ ★
Best Seen:	Winter

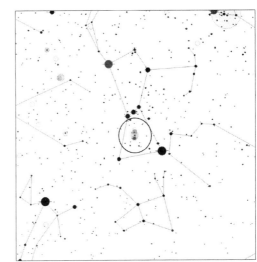

Map courtesy *Mobile Observatory*

By far the brightest and best nebula visible from the northern hemisphere, the Orion Nebula can be described with a single word: stunning.

Easily seen with just the unaided eye, it'll look good even through binoculars or a halfway decent finderscope. The cloud appears small and misty, but you should be able to pick out three or four bright stars in the heart of the nebula. These stars, known as the Trapezium, are young stars recently born from the nebula and still swaddled in the surrounding stellar nursery.

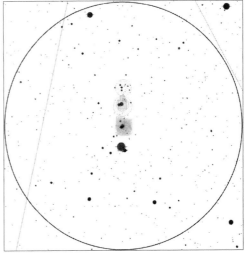

Finderscope view courtesy *Mobile Observatory*

Although best observed with low or medium magnification, the nebula stands up to quite a bit of power with greater magnifications revealing more details to the observer. Unlike some celestial sights, you'll notice some color here too – observers often report a greenish tint with the nebula appearing smoke-like. I've noted that it looks as though it's illuminated by moonlight.

While you're here, look out for a small, circular misty patch nearby with a lone star within it. This is Messier 43 (M43), a fragment that appears separated from the main nebula itself. You'll also see a darker, triangular shape on that edge, close to the Trapezium and M43. This is known as the Fish's Mouth and appears as a result of darker matter that obscures the light of the nebula beyond.

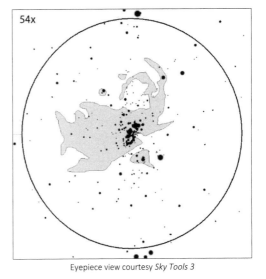

Eyepiece view courtesy *Sky Tools 3*

The Orion Nebula is about 1,300 light years away and some 12 light years in diameter.

Sigma Orionis

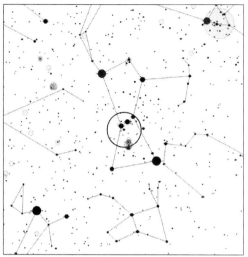

Map courtesy *Mobile Observatory*

Designation(s):	Sigma Orionis
Constellation:	Orion
R.A.:	05h 39m 34s
Declination:	-02° 35' 36"
Object Type:	Multiple Star
Location:	★ ★ ★
Rating:	★ ★ ★
Best Seen:	Winter

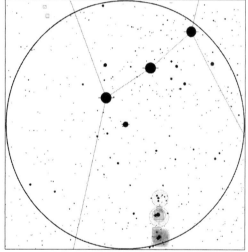

Finderscope view courtesy *Mobile Observatory*

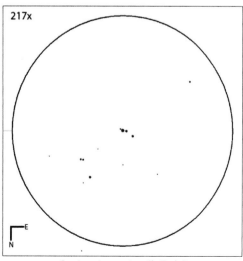

Eyepiece view courtesy *Sky Tools 3*

Sigma is quite a fascinating multiple star that's easily found just below Orion's belt. What's particularly interesting is that it appears close to a mini belt at 26x.

I've also noted that it appears in the middle of an asterism that resembles the constellation Sagitta, with two stars to the west that form the feathers of the arrow. The southernmost star is itself a double with both components being of equal brightness.

Sigma itself has three components with the primary being a brilliant white star. There is a secondary, wide companion to the north-east that appears to be white and about 1½ times fainter than the primary. Look out for a slightly fainter third star, much closer than the second, which is also white and just east of the primary.

The system lies about 1,150 light years away and actually comprises five stars. The primary you see in a telescope is a very close pair of young white dwarves that orbit one another once every 170 years.

The next star in the system is just to the west of the primary but will be too close and faint to be seen with small scopes from suburban or city skies.

The remaining two stars are also dwarves about seven times the mass of the Sun with one being a little-understood, helium-rich magnetic star.

Gamma Leporis

Designation(s):	Gamma Leporis
Constellation:	Lepus
R.A.:	05h 44m 28s
Declination:	-22° 26' 54"
Object Type:	Multiple Star
Location:	★ ★ ★
Rating:	★ ★
Best Seen:	Winter

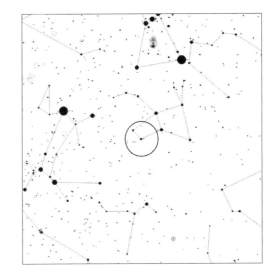

Map courtesy *Mobile Observatory*

Also split with a regular pair of 10x50 binoculars, this is a fairly wide double that should reveal both the two brightest components at a low magnification.

At 26x, the primary appears to be a creamy white and about twice as bright as the coppery-orange secondary. (On one occasion, I noted the companion appeared to have a blue-purple tint.)

Things get a little more interesting when you up the power. If you increase the magnification to about 90x, you might be able to see a third, fainter star, flickering in and out with averted vision. It appears about three times further away than secondary.

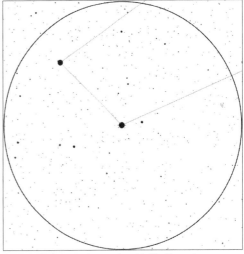

Finderscope view courtesy *Mobile Observatory*

Go a little further – about 100x or even a little higher – and you might catch a fourth star. It's very faint and appears on the opposite side of the primary from the second star. If you look at the distance between the primary and secondary stars, the fourth should appear a little further away.

At just under thirty light years, this system is relatively close to the Earth and is interesting for one other major reason: the primary star is Sun-like (although only about half the age) and may have Earth-sized planets orbiting it. Something to consider as you stare across space at this fascinating star.

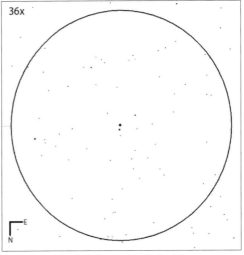

Eyepiece view courtesy *Sky Tools 3*

Messier 37

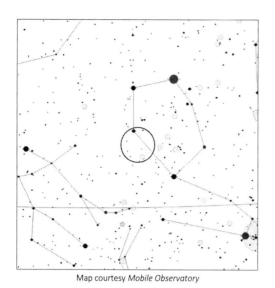

Map courtesy *Mobile Observatory*

Designation(s):	Messier 37
Constellation:	Auriga
R.A.:	05h 52m 18s
Declination:	+32° 33′ 02″
Object Type:	Open Cluster
Location:	★
Rating:	★ ★ ★
Best Seen:	Winter

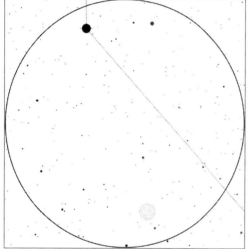

Finderscope view courtesy *Mobile Observatory*

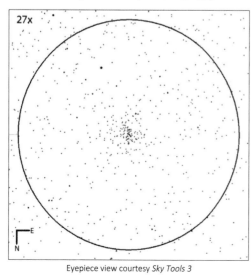

Eyepiece view courtesy *Sky Tools 3*

Messier 37 is the brightest of three Messier open star clusters in Auriga and, lying close to Theta Aurigae, is also the easiest found.

With Theta on the edge of your finderscope field of view, look for the cluster close to the opposite edge. It should appear as a very faint, misty patch under suburban skies.

The other two clusters, Messier 36 and Messier 38, lie to the west, in the south-eastern portion of Auriga's misshapen hexagon and are also worth seeking out.

It's a rich cluster (the richest of the three) and should be easily observed from suburban skies but may be more problematic from the city.

It presents a very nice view at 35x and I noted that there seemed to be a thousand or more stars in the cluster. Using averted vision, it seemed as though a thousand more would spring into view. I've also noted a number of triangular patterns within the cluster. You should be able to increase the magnification to about 50x and still fit the cluster within the field of view.

It's definitely a cluster that begs to be stared at. Just allow your eyes to take in the view, enjoy it and consider that it lies about 4,500 light years away and its 500 stars are scattered across some 25 light years of space.

Designation(s):	Messier 35
Constellation:	Gemini
R.A.:	06h 08m 56s
Declination:	+24° 21' 28"
Object Type:	Open Cluster
Location:	★★
Rating:	★★★
Best Seen:	Winter

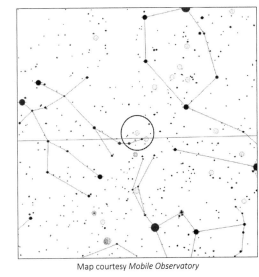

Map courtesy *Mobile Observatory*

Messier 35 is a large, bright open star cluster that should be easily found with your finderscope – or even with just your unaided eyes.

Through binoculars, it appears to be a faint, grey misty patch of uniform brightness but with an hourglass shape. If you look with averted vision, the patch may appear to be resolved into hundreds of tiny points of light – the cluster's individual stars.

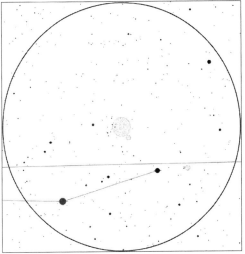

Finderscope view courtesy *Mobile Observatory*

Given its size, this is a cluster that's best observed at lower power and about 30x should do the trick. At this magnification, the cluster nicely fits into the field of view and appears to be surrounded by a background scattering of stars.

It's fairly sparse but with plenty of faint stars. Look for a pale gold star on the edge of the cluster with a blue star close to it, giving it the impression of a double star. The golden star is about 1½ times brighter than its blue companion.

Of the others, the remaining stars appear to be either blue or blue-white.

The cluster was discovered in 1745 and lies at a distance of 2,800 light years from Earth.

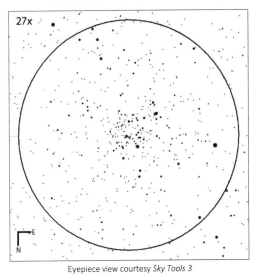

Eyepiece view courtesy *Sky Tools 3*

Castor

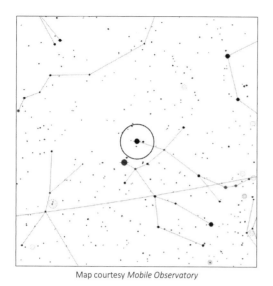

Map courtesy *Mobile Observatory*

Designation(s):	Alpha Geminorum
Constellation:	Gemini
R.A.:	07h 34m 36s
Declination:	+31° 53' 18"
Object Type:	Multiple Star
Location:	★ ★ ★
Rating:	★ ★ ★
Best Seen:	Winter

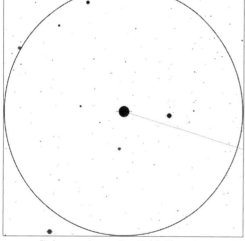

Finderscope view courtesy *Mobile Observatory*

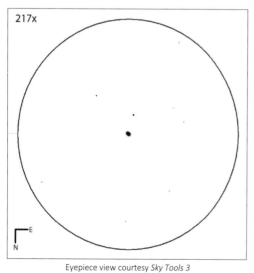

Eyepiece view courtesy *Sky Tools 3*

Castor, despite being designated as Alpha Geminorum, is actually the second brightest star in the constellation of Gemini, the Twins. To the unaided eye, it appears to be a solitary, bright white star but in reality it's a quadruple star system located some 51 light years away.

Unfortunately, only two of the four can be seen with amateur equipment and a magnification of about 50x is needed to split it.

At 26x you'll probably notice two very faint companions of almost equal magnitude within the same field of view. Although this is interesting in itself, this isn't what you're here to see.

Try upping the magnification to about 50x and take another look. Does Castor appear to be split? On some nights, I've barely split it at 54x. On other occasions, I can't say it's been resolved. A lot will depend upon your equipment and the sky conditions.

Double the magnification to about 100x and the view improves immensely. Castor should now be closely split into a pair of brilliant white stars, of almost equal magnitude. Through the 130mm scope, there also appeared three fainter stars visible nearby, equally spaced out like Orion's belt but with the middle star a little out of line from the others.

Don't miss this winter wonder!

Designation(s):	Beta Monocerotis
Constellation:	Monoceros
R.A.:	06h 28m 49s
Declination:	-07° 01' 59"
Object Type:	Multiple Star
Location:	★ ★
Rating:	★ ★
Best Seen:	Winter

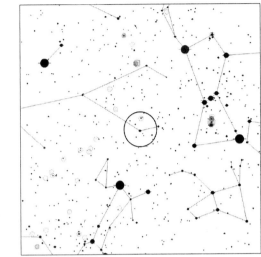

Map courtesy *Mobile Observatory*

Beta Monoceros is a true triple star system some 700 hundred light years away. Two of the stars should be a relatively easy target for a small scope and a low magnification eyepiece.

It's barely split at 26x but boosting the magnification to 35x or more should make it cleaner and more obvious.

You'll see a pair of white stars of equal brightness with a third, fainter star nearby. (My estimates of the third star's brightness seem to vary somewhat.)

Increasing the magnification might reveal some color (pale yellow and pale blue,) but I only noticed this from the city so there's a real chance it had more to do with the quality of the air than the stars themselves.

You may need to push your scope to the limit to see all three stars. If you have a Barlow, increase your magnification to as far as your scope can usefully go and try your luck.

At 217x, with the 150mm, I noted that one of the two brightest stars appeared to be elongated and, therefore, potentially multiple, but this was at the limit of the telescope's maximum magnification and I was unable to actually split them.

What do you see?

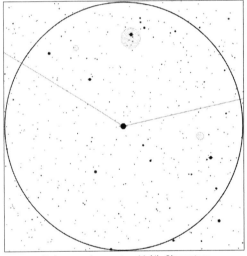

Finderscope view courtesy *Mobile Observatory*

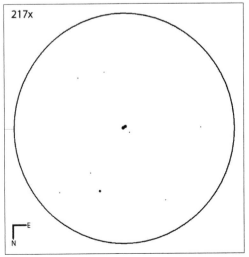

217x

Eyepiece view courtesy *Sky Tools 3*

Sirius

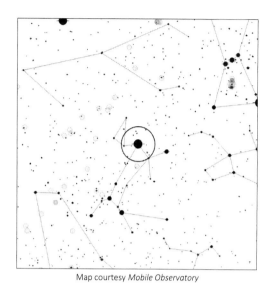

Map courtesy *Mobile Observatory*

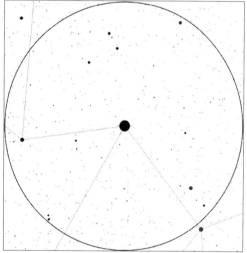

Finderscope view courtesy *Mobile Observatory*

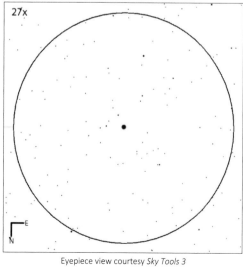

Eyepiece view courtesy *Sky Tools 3*

Designation(s):	Alpha Canis Majoris
Constellation:	Canis Major
R.A.:	06h 45m 09s
Declination:	-16° 42' 58"
Object Type:	Multiple Star
Location:	★ ★ ★
Rating:	★
Best Seen:	Winter

Sirius, as many people know, is the brightest star in the entire night sky and has been known throughout history by many different civilizations and cultures.

It's a brilliant blue-white star that will often sparkle and flash a myriad of colors when close to the horizon. In fact, its name is actually derived from the Greek for "glowing" and it's not unusual for unsuspecting folk to report it as a UFO.

Easily found by drawing a line through Orion's belt down and toward the east, it's a very easy target and you may have already turned your telescope toward it. What did you see? Personally, I was very surprised to find that it can be dazzling when observed, even at low power.

At 26x I've noticed that it forms a triangle with two much fainter stars and there's a fourth, coppery star between those two that's fainter than all of them.

Sirius is a little over eight and a half light years away and is one of the closest stars to the Earth. Besides the primary star itself, there's a smaller white dwarf companion that can be glimpsed with large telescopes.

This companion was once a red giant that slowly died and shrunk about 120 million years ago. As the primary, Sirius A, is often known as the "dog star" the white dwarf companion, Sirius B, is sometimes nicknamed "the pup."

Designation(s):	Messier 41
Constellation:	Canis Major
R.A.:	06h 46m 00s
Declination:	-20° 45' 15"
Object Type:	Open Cluster
Location:	★ ★
Rating:	★ ★ ★
Best Seen:	Winter

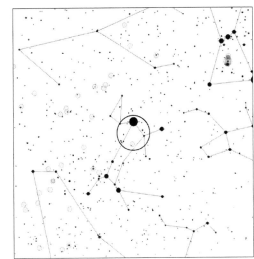

Map courtesy *Mobile Observatory*

Messier 41 is one of my favorite clusters, for a number of reasons. Firstly, it's very easily located, thanks to the proximity of Sirius. All you need to do is point your finder toward that bright star and with Sirius near the top, the cluster will appear near the bottom.

Through binoculars under suburban skies, it appears quite large and has a definite shape to it. It looks as though there's a circular hole on the western edge of the cluster that gives it the appearance of a lobster with its claws outstretched.

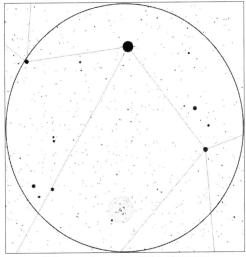

Finderscope view courtesy *Mobile Observatory*

That impression is reinforced when the cluster is observed at 35x. I feel it's one of the best (and under-rated) clusters in the sky, one that allows your imagination to run wild.

It's a nice, large cluster with sparsely scattered stars that reminded me of NGC 457 in Cassiopeia, only with a lot more stars. The vast majority are white or blue-white and many are of about the same brightness. However, there's a pair of stars, just a little brighter than the others and another with an orange hue. This reminds me of Phi Cassiopeiae.

The other stars in the cluster are grouped - to me - as though they're forming the shape of a man with his hands up, or the aforementioned lobster... or wings like an angel... or a dove... what do you see?

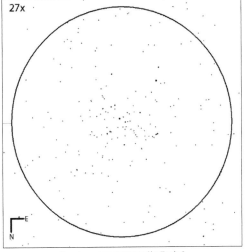

Eyepiece view courtesy *Sky Tools 3*

Procyon

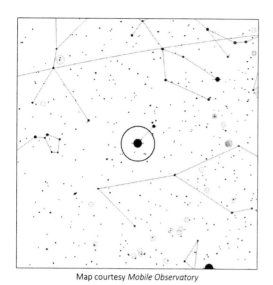

Map courtesy *Mobile Observatory*

Designation(s):	Alpha Canis Minoris
Constellation:	Canis Minor
R.A.:	07h 39m 18s
Declination:	+05° 13' 29"
Object Type:	Multiple Star
Location:	★ ★ ★
Rating:	★
Best Seen:	Winter

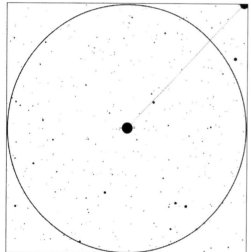

Finderscope view courtesy *Mobile Observatory*

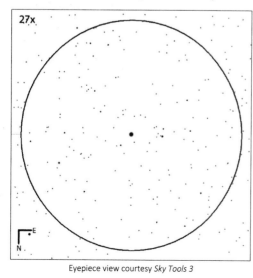

Eyepiece view courtesy *Sky Tools 3*

Procyon is a bright star, easily located to the east of Orion. Use the two stars of his shoulders - Betelgeuse and Bellatrix - to help identify it. (See page 37.)

Through the finderscope (or binoculars) it appears as a brilliant white star with a wide, faint blue-white companion nearby. This companion appears about ten times fainter than Procyon itself.

The pair will still appear within the same field of view at 35x but you'll also notice a scattering of many fainter stars in the background. In particular, look out for another faint star near Procyon and two others near the companion.

Procyon is the brightest of only two bright stars in the constellation of Canis Minor, the Little Dog. As with Sirius, the brightest star in Canis Major, the Great Dog, it follows Orion across the sky, just as the two dogs follow their master as he embarks on his winter hunting trips.

(They all seem totally oblivious to their prey, Lepus the Hare, which can be found below Orion. Maybe they've been distracted by Taurus the Bull, charging from the west.)

Again, like Sirius, Procyon is another bright, white star with a white dwarf companion that's just under eleven and a half light years away. The pair orbit one another every forty years.

Designation(s):	Messier 93
Constellation:	Puppis
R.A.:	07h 44m 29s
Declination:	-23° 51' 11"
Object Type:	Open Cluster
Location:	★
Rating:	★ ★
Best Seen:	Winter

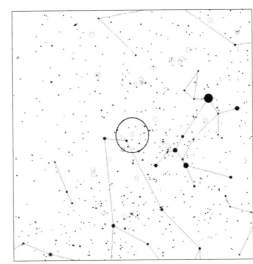

Map courtesy *Mobile Observatory*

This open star cluster is quite easily found, just to the north-west of Xi Puppis. It's a medium-sized cluster, quite dense with the vast majority of stars being blue-white with many other fainter stars visible with averted vision.

To get the best view, you'll want to get away from the city as I could barely see it at 26x. At that magnification, it appeared small but I was still able to see two bright stars at the tip of a group of much fainter stars. The group looked like a curved triangle.

The view from darkened skies is much better as many of the fainter stars are more easily visible. At 35x, the cluster made me think of a butterfly with the two bright, orange stars on the tip of the western wing.

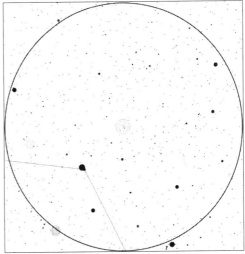

Finderscope view courtesy *Mobile Observatory*

Whether you're observing from the suburbs, the country or from the city itself, the cluster stands up well to higher magnification. I found the best view to be had at around 75x. Increasing the magnification to about 100x only made the cluster appear larger, rather than richer. (A larger telescope, with greater light gathering power, would almost certainly reveal more stars.)

The cluster was first recorded by Charles Messier in March 1781, just a week after William Herschel discovered the planet Uranus. The cluster lies about 3,600 light years away and is thought to contain about 80 stars and is about 25 light years in diameter.

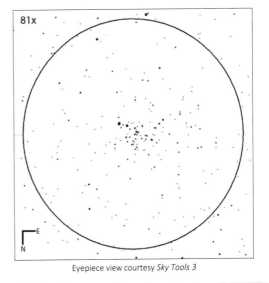

Eyepiece view courtesy *Sky Tools 3*

The Praesepe

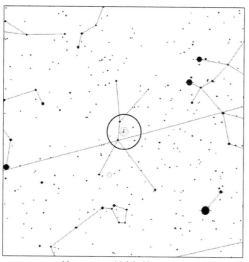

Map courtesy *Mobile Observatory*

Designation(s):	Messier 44
Constellation:	Cancer
R.A.:	08h 40m 22s
Declination:	+19° 40' 19"
Object Type:	Open Cluster
Location:	★ ★
Rating:	★ ★ ★
Best Seen:	Winter and Spring

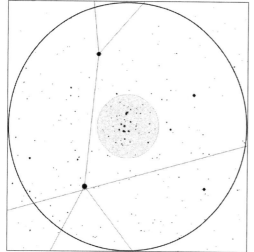

Finderscope view courtesy *Mobile Observatory*

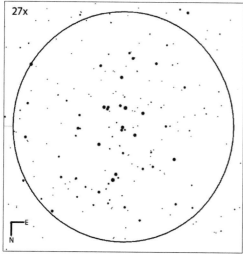

Eyepiece view courtesy *Sky Tools 3*

Known since antiquity, the traditional name for the cluster is the Praesepe, which means "the manger." Another popular name is the Beehive Cluster. Personally, it always gives me the impression of a swarm of bees, but I also like the traditional name.

Whatever name you choose, you'll find spotting the cluster to be a good test of your environment. In theory, you should be able to see it with just your eyes but finding it can be a little problematic as it's actually brighter than any of the stars that form the constellation of Cancer itself.

Look midway between the mid-section of Gemini and Regulus, the brightest star in Leo, the Lion. With luck, you should be able to see a tiny, misty patch. I've seen it from the suburbs but in the city I've always needed a pair of binoculars to help me.

It's a large cluster and, consequently, one that's best suited to lower magnifications. Even at 35x it'll barely fit within the field of view. But it's a very nice view, all the same with dozens of stars glinting against the night. The majority of stars appear blue-white, indicating a young age. Several others are older orange giants.

There are a lot of doubles to be found here, but look toward the center: can you see a mini Cepheus? It always leaps right out at me!

Designation(s):	Iota Cancri
Constellation:	Cancer
R.A.:	08h 46m 42s
Declination:	+28° 45' 36"
Object Type:	Multiple Star
Location:	★ ★ ★
Rating:	★ ★ ★
Best Seen:	Winter and Spring

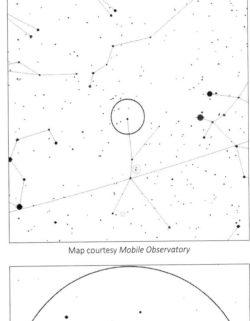

Map courtesy *Mobile Observatory*

Iota Cancri is like a pale Albireo for late winter and early spring. Unfortunately, it's not a prominent star, which means you might have difficulty finding it from the city. Scan the area about midway between Pollux and the top of the backwards question-mark that represents Leo's head.

Through a finderscope or binoculars, you'll see 53 and Rho Cancri - another, wider coppery double pair to the east but within the same field of view. This is a nice sight for binoculars and may be a consolation prize for anyone observing without a telescope.

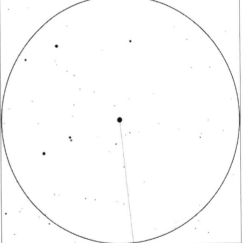

Finderscope view courtesy *Mobile Observatory*

Once you've found Iota, you'll find it easy to split with a low powered eyepiece. Just 26x will reveal both components. The primary is pale gold and appears to be about two or three times brighter than the very pale blue secondary.

Although it looks good at all magnifications, it's probably best at about 100x where a decent distance appears between the pair. As a bonus, I've noted that the secondary shows definite signs of purple and deep blue when observed at higher powers.

The pair is not a true multiple star system – it's what's known as an *optical* binary. In other words, the stars only appear close to another due to chance alignment.

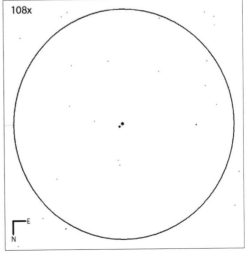

Eyepiece view courtesy *Sky Tools 3*

Messier 67

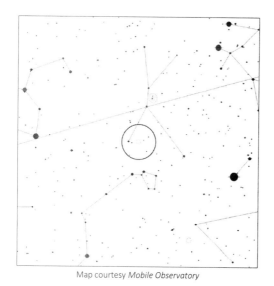

Map courtesy *Mobile Observatory*

Designation(s):	Messier 67
Constellation:	Cancer
R.A.:	08h 51m 20s
Declination:	+11° 48' 43"
Object Type:	Open Cluster
Location:	★
Rating:	★ ★
Best Seen:	Winter and Spring

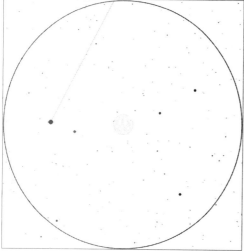

Finderscope view courtesy *Mobile Observatory*

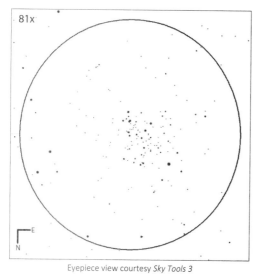

Eyepiece view courtesy *Sky Tools 3*

This late Winter / early Spring cluster is reminiscent of Messier 37 in Auriga. It's a tightly compacted cluster of faint blue-white stars that mostly appear to be about the same magnitude. There are a couple of notable exceptions, however.

It's barely visible from the city but can be glimpsed at 26x and using averted vision will help to bring out the faintest stars. In particular, look for a bright orange star that appears on the edge of the shield-shaped cluster. You should also see a pair of faint stars on the opposite edge but the remaining stars might be difficult.

Increasing the magnification will help – it will still fit within the field of view at about 90x - but the best views are reserved for those observing from a dark location.

Even at 35x, you'll see a glimmering diamond shape that appears to contain hundreds of tiny stars. To me, again, like Messier 37, it gave me the impression of sugar or even diamond dust on velvet but besides the overall shape of the cluster itself I didn't notice any particular patterns.

At more than three billion years old the cluster is one of the oldest known and lies at a distance of over 2,500 light years. Containing more than five hundred stars, it's thought to be about ten light years in diameter.

Designation(s):	Alpha Leonis
Constellation:	Leo
R.A.:	10h 08m 22s
Declination:	+11° 58' 02"
Object Type:	Multiple Star
Location:	★ ★ ★
Rating:	★
Best Seen:	Spring

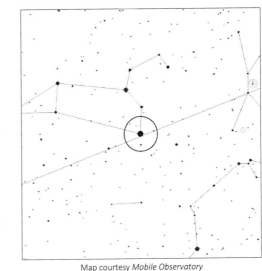
Map courtesy *Mobile Observatory*

Regulus is one of the brightest stars in the sky and is also one of the most overlooked – at least by experienced amateurs.

Through binoculars or a finderscope, you'll see the reasonably bright stars Nu and 31 Leonis within the same field of view, but look carefully with binoculars and you might see something else.

You'll need to hold your binoculars steady and you may have to wait for the air to be still but you might just see a very faint companion to the north-west. (If your finder magnifies the view, try your luck with that too.)

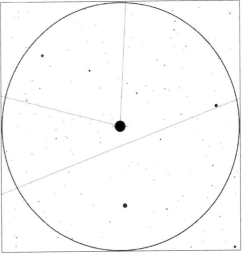
Finderscope view courtesy *Mobile Observatory*

The pair are much more apparent at low power in a small telescope. At 27x, you'll see Regulus as a brilliant white star, much brighter than the bluish secondary. It'll be an easy and wide split but if you want the secondary to be more easily seen, you'll need to up the power some more. I found that 50x provided quite a nice view.

Regulus is the brightest star in the constellation of Leo the Lion and, at 79 light years away, is relatively close to Earth. The pair are a part of a true multiple star system with both components orbiting one another once every two million years.

Each component is itself double, making this a four star system in all.

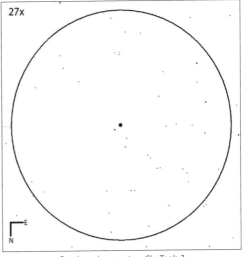

27x

Eyepiece view courtesy *Sky Tools 3*

Adhafera

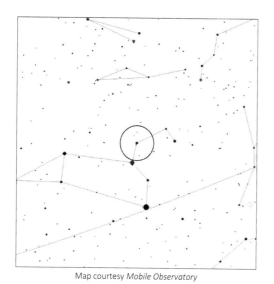

Map courtesy *Mobile Observatory*

Designation(s):	Zeta Leonis
Constellation:	Leo
R.A.:	10h 16m 41s
Declination:	+23° 25′ 02″
Object Type:	Multiple Star
Location:	★ ★ ★
Rating:	★ ★
Best Seen:	Spring

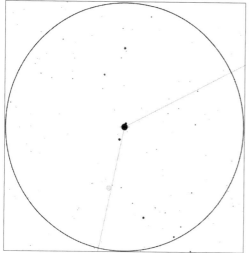

Finderscope view courtesy *Mobile Observatory*

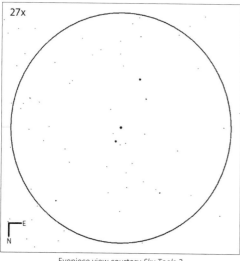

27x

Eyepiece view courtesy *Sky Tools 3*

Adhafera is another overlooked star in Leo, the Lion. Through the finder or binoculars you'll see a fainter star, 35 Leonis, to the south. This star is not associated with Adhafera and its position is purely due to chance.

Specifically, Adhafera is about 260 light years away while 35 Leonis is about 100 and both stars are moving in different directions.

Unfortunately, I don't have any notes from binocular observations, but you may also be able to spot another, fainter companion to the north-west. It'll appear very close to Adhafera itself with the pair best observed through a small telescope.

At 27x you'll see a wide pair of stars comprising of a lemony-white primary that's about 1½ times brighter than its white companion.

As with many double stars, I've noted some variance in the colors; sometimes the primary appears to be purely white while the secondary can appear peachy.

Adhafera is notable for another important reason – the Leonid meteors appear to radiate from a point just to the west of the star. This shower peaks around November 18th and usually produces about ten or twenty shooting stars per hour. However, on occasion, the shower has been known to exceed a hundred.

Designation(s):	Gamma Leonis
Constellation:	Leo
R.A.:	10h 19m 58s
Declination:	+19° 50' 30"
Object Type:	Multiple Star
Location:	★ ★ ★
Rating:	★ ★ ★
Best Seen:	Spring

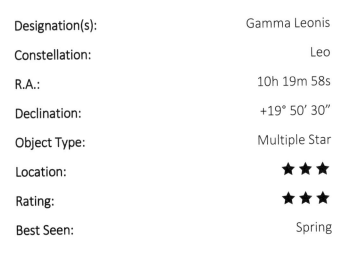

Map courtesy *Mobile Observatory*

Algieba is an easy, wide double star that can be easily seen with either binoculars or even your own unaided eyes. The primary appears gold and about two or three times as bright as the bluish companion. You might also notice a third, slightly fainter star that forms a triangle with the two brighter stars.

The real magic happens when you turn a telescope toward the trio. You won't notice anything different until you get to about 50x or higher. The gold star may be barely split at this magnification with a pale white gold companion appearing close beside it.

If you're having difficulty with the pair, it might be due to your location and/or your observing conditions. I've been able to comfortably split it at under 100x from the suburbs but from the city it's barely split at even 162x. (This, no doubt, is also due to the higher humidity of my location and the fact that I was observing the star in June, when temperatures were still 70°F at night.)

What's the lowest magnification you can use to split the star and see the secondary? Find a magnification where the secondary is clearly seen and then trying working your way back down until it disappears again. At what point does Algieba go from being a clean split to elongated and then just a single point of light?

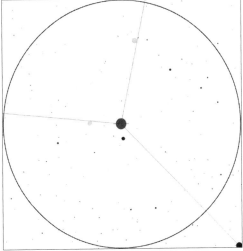

Finderscope view courtesy *Mobile Observatory*

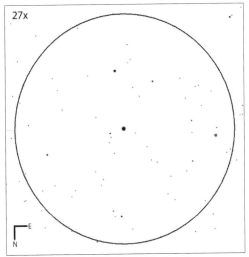

27x

Eyepiece view courtesy *Sky Tools 3*

Denebola

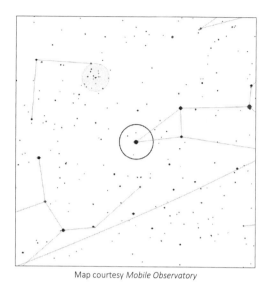

Map courtesy *Mobile Observatory*

Designation(s):	Beta Leonis
Constellation:	Leo
R.A.:	11h 49m 04s
Declination:	+14° 34' 19"
Object Type:	Multiple Star
Location:	★ ★ ★
Rating:	★ ★
Best Seen:	Spring

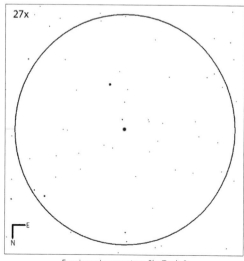

Finderscope view courtesy *Mobile Observatory*

Eyepiece view courtesy *Sky Tools 3*

Denebola is a bright star that marks the tail of Leo the Lion – and it's not one you usually find highlighted in an astronomy book.

In fact, I really didn't pay any attention to it until I bought a small 70mm reflector for my son and was looking for easy objects for him to observe. In essence, it was the beginnings of the very book you're reading now and demonstrates the benefits of going off the beaten track and sometimes just randomly exploring the night sky.

Through binoculars or a finderscope, you'll see a wide pair of stars, similar to Algieba. Both are white with the primary about four times brighter than the secondary.

Similarly, if you observe the star through a small telescope – even at a low magnification of 15x – you'll easily see the pair but a third, much fainter star will become apparent. It's a lot fainter than the primary and can be seen about a quarter of the way between the first two stars. You may need to use averted vision to see it properly.

I then decided to take a look through my 130mm scope and increased the magnification even further. Once I got to 108x, I got another surprise. While all three stars could still be seen within the field of view, a fourth star now appeared, fainter than the third. Look for it about midway between the first two and a little out of line with the others.

Designation(s):	Delta Corvi
Constellation:	Corvus
R.A.:	12h 29m 52s
Declination:	-16° 30' 56"
Object Type:	Multiple Star
Location:	★ ★ ★
Rating:	★ ★
Best Seen:	Spring

Map courtesy *Mobile Observatory*

Algorab is the third brightest star in the constellation of Corvus the Crow. In Greek mythology, the god Apollo sent the crow to fetch him a cup of water but the bird stopped to eat figs and was slow to return. To punish the bird, Apollo threw him and the cup into the sky. Consequently, Crater, the Cup, appears to the west of Corvus and will forever be out of reach of the bird.

Through a finder or binoculars, you'll see a wide, unrelated companion star, just to the north-east. You may even be able to see it with just your eyes. However, you'll need a telescope to see both components of Algorab itself.

The star was usually split at 26x but it was dependent upon the quality of the atmosphere, especially from the city. It can be a clean split but the secondary is very faint and may not be seen unless you use a higher magnification first.

It was a lot easier under suburban skies at 35x. The primary appeared white with a very pale and faint rusty colored secondary close beside it. Increasing the magnification might improve the view but I've noticed the colors appear to fade in doing so.

In reality, the two stars are moving along a similar path through space but they're not thought to be a true multiple star system.

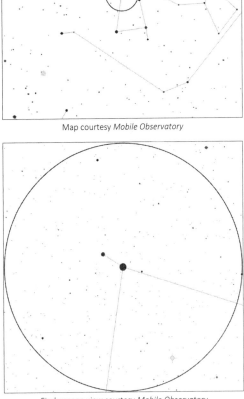

Finderscope view courtesy *Mobile Observatory*

108x

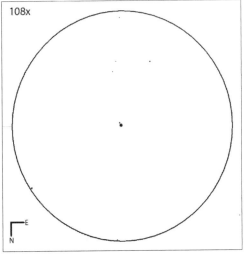

Eyepiece view courtesy *Sky Tools 3*

Cor Caroli

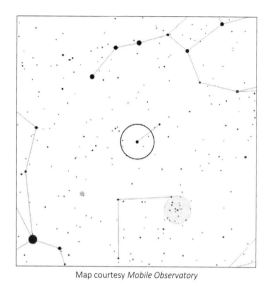

Map courtesy *Mobile Observatory*

Designation(s):	Alpha Canum Venaticorum
Constellation:	Canes Venatici
R.A.:	12h 56m 02s
Declination:	+38° 19' 06"
Object Type:	Multiple Star
Location:	★ ★ ★
Rating:	★ ★ ★
Best Seen:	Spring and Summer

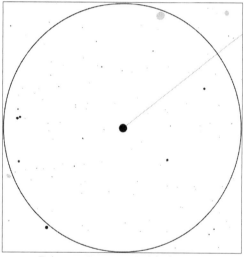

Finderscope view courtesy *Mobile Observatory*

The name Cor Caroli actually means "Charles's Heart" and it originates from the 17ᵗʰ century when it was said the star shone particularly brightly upon the King's return to England.

It's a showpiece multiple star for the Spring and Summer skies and another favorite with astronomers throughout the northern hemisphere. Even at a low magnification of just 26x the star is clearly split and provides a beautiful view. Increasing the power to about 50x or 60x will probably give you the best view.

The primary is white and appears about three times brighter than the secondary. However, I've noted slightly different colors for the secondary; more often than not, I've noted that it appears white too, but on occasion it seems to hold a creamy or pale gold color. (I've even noted a grey-greenish color at a magnification of 182x.)

What colors do you see?

(Most other sources describe the pair as both being either white or blue-white in color but some have also noted the different colors reported by observers.)

Cor Caroli is a true binary star system some 110 light years away with the pair orbiting one another roughly every 8,300 years. Curiously, the primary component is highly magnetic and has a magnetic field thousands of times stronger than the Earth's.

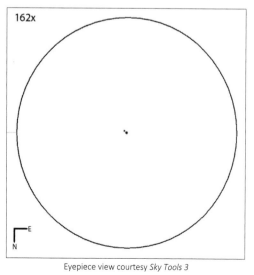

Eyepiece view courtesy *Sky Tools 3*

Mizar and Alcor

Designation(s):	Zeta and 80 Ursae Majoris
Constellation:	Ursa Major
R.A.:	13h 23m 56s
Declination:	+54° 55' 31"
Object Type:	Multiple Star
Location:	★ ★ ★
Rating:	★ ★ ★
Best Seen:	Spring and Summer

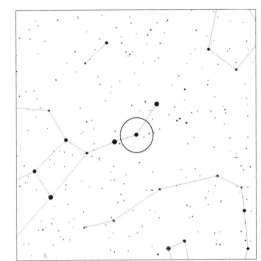

Map courtesy *Mobile Observatory*

Mizar and Alcor are a wide pair of stars that have been considered a test of eyesight since ancient times. As both stars are well within the limits of naked eye visibility, I've never had any trouble seeing them both – even from the city.

If you're familiar with the seven stars of the Dipper (or the Plough, as it's known in the United Kingdom) you'll find the pair in the middle of the curved line of three that make up the handle. Mizar is, of course, immediately visible but you might need to stare for a few moments to spot Alcor.

The pair make for an attractive sight in binoculars with Mizar being about twice as bright as Alcor and both stars appearing blue-white.

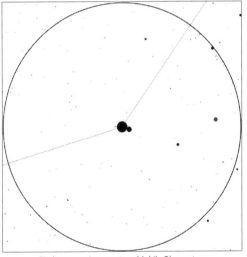

Finderscope view courtesy *Mobile Observatory*

However, like many other doubles, it's only when you turn a telescope toward the stars that everything comes into view. At a low power of just 27x, Mizar is clearly and easily split with the secondary also blue-white and about twice as faint as the primary.

While you're in the area, look out for a third star that forms a triangle with Alcor and the two components of Mizar.

Mizar and Alcor are not a true double star system but Mizar itself is a quadruple star some 85 light years away.

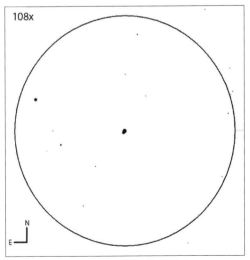

Eyepiece view courtesy *Sky Tools 3*

Arcturus

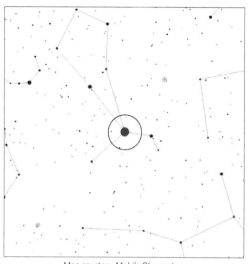

Map courtesy *Mobile Observatory*

Designation(s):	Alpha Boötis
Constellation:	Boötes
R.A.:	14h 15m 40s
Declination:	+19° 10' 56"
Object Type:	Multiple Star
Location:	★ ★ ★
Rating:	★
Best Seen:	Spring and Summer

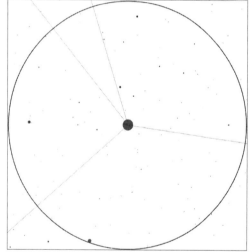

Finderscope view courtesy *Mobile Observatory*

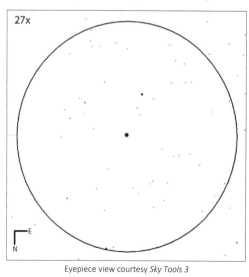

Eyepiece view courtesy *Sky Tools 3*

Arcturus is the fourth brightest star in the sky and, at just under 37 light years away, one of the closest to the Earth.

It's a red giant star that appears orange to the unaided eye but despite this, at 27x I've noted that it shows a brilliant, white gold hue. Other observers have noted a range of colors including yellow, peach and topaz.

(If you want to see the star truly sparkle, wait until it's low on the horizon. Like Sirius, you'll find it flashing a multitude of colors when it's close to setting in the west.)

This has always been something of a favorite star for me. It's easily found by following the curved tail of Ursa Major, the Great Bear, down towards the south. ("Arc down to Arcturus.") It appears at the bottom of kite-shaped Boötes, a large constellation representing a herdsman of the same name. (The name Arcturus actually means "guardian of the bear.")

Arcturus is an old star, maybe seven billion years old, but is still over a hundred times more luminous than the Sun. As one of the fastest moving stars, it's currently drawing nearer and will be at its closest in about 4,000 years. In the far more distant future, it will eventually shed its outer layers and shrink down to a white dwarf, forming a planetary nebula in the process. Unfortunately, that sight will only be seen by our descendants as you and I will both be long gone!

Designation(s):	Delta Boötis
Constellation:	Boötes
R.A.:	15h 15m 30s
Declination:	+33° 18' 53"
Object Type:	Multiple Star
Location:	★ ★ ★
Rating:	★ ★ ★
Best Seen:	Spring and Summer

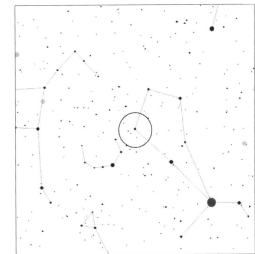

Map courtesy *Mobile Observatory*

There are a number of multiple stars to be found in Boötes but Delta Boötis is one of the easiest to find. It's a relatively bright star that marks the upper-left corner of the kite-shaped constellation.

Although a fairly wide double, I haven't been able to split the pair with binoculars but others have noted that it's been done so it's definitely worth a try. (I was using a smaller pair of 8x30's so a standard pair of 10x50's might be just enough to split them.)

If you're using a small telescope, a low magnification of about 25x should easily do the trick.

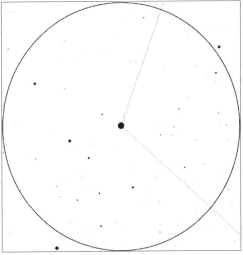

Finderscope view courtesy *Mobile Observatory*

At that magnification, it appears to be a wide and pretty pair with a nice color contrast between the two stars. The pale yellow-white primary appears about five times brighter than pale blue secondary. Add to that a number of other stars that appear within the same field of view and you have a sight worth looking out for.

Delta is thought to be a true double star system, lying some 121 light years from Earth. The primary is a giant star nearing the end of its life and some ten times larger than our own Sun while the secondary is a dwarf star a little smaller than the Sun. The pair are thought to orbit one another once every 120,000 years.

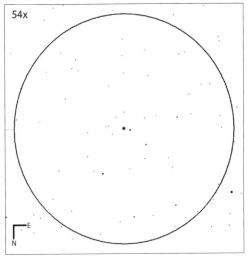

Eyepiece view courtesy *Sky Tools 3*

Zuben Elgenubi

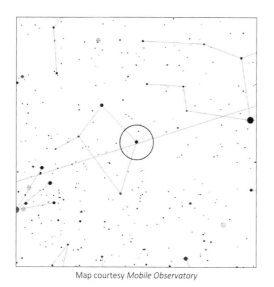

Map courtesy *Mobile Observatory*

Designation(s):	Alpha Librae
Constellation:	Libra
R.A.:	14h 50m 41s
Declination:	-15° 59' 50"
Object Type:	Multiple Star
Location:	★ ★ ★
Rating:	★ ★
Best Seen:	Spring

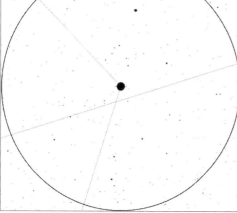
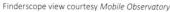

Finderscope view courtesy *Mobile Observatory*

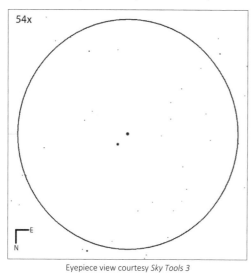

Eyepiece view courtesy *Sky Tools 3*

Zuben Elgenubi – or Alpha Librae, if you prefer – is a wide pair of stars easily seen with both binoculars and telescopes at low power. In fact, this is one example where the object is best at low power as increasing the magnification too much reduces the aesthetic appeal of the stars.

Both stars are white but the primary is about 1½ times brighter than the secondary.

The name Zuben Elgenubi is Arabic and means "the southern claw." In case you're wondering how that relates to a constellation that represents scales, it originates from a time when the stars of Libra were still considered part of nearby Scorpius, the Scorpion.

Despite it having the designation of Alpha Librae, it is in fact the second brightest star in the constellation. It's thought to be a true binary system (as opposed to being just a chance alignment of two stars) that lies about 77 light years away. The pair are estimated to obit one another once every 200,000 years.

Each star is itself a binary system, making four components in all with a suspected fifth member – KU Librae – appearing nearby.

Designation(s):	Beta Scorpii
Constellation:	Scorpius
R.A.:	16h 05m 26s
Declination:	-19° 48' 20"
Object Type:	Multiple Star
Location:	★ ★ ★
Rating:	★ ★ ★
Best Seen:	Spring and Summer

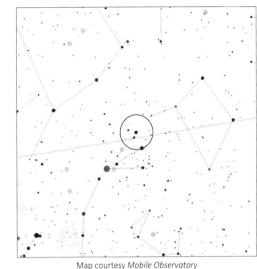

Map courtesy *Mobile Observatory*

Graffias is an interesting star – and also slightly confusing as the name applies to both Beta and Xi Scorpii. To clarify, Beta appears in the center of the finderscope depiction to the right and is the star we're interested in.

If you're using binoculars, you should easily see Omega[1] and Omega[2], a fairly wide pair of white stars nearby. These are the two stars that can be seen off center, just to the lower left of Beta in the finderscope view. You might also notice a couple of other fainter stars in the field of view.

This in itself is a nice enough view, but if you turn your telescope toward Beta you should be able to split it at 27x without any trouble.

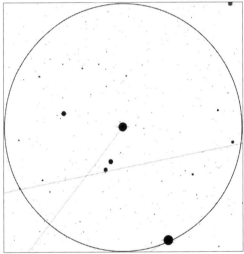

Finderscope view courtesy *Mobile Observatory*

Both components of the primary appear white (or sometimes blue-white) with the secondary being about three times fainter than the primary. It's a pretty nice view, even at this magnification.

Look carefully at Beta's secondary as I've noticed it showed signs of a violet color at 81x. However, this was while observing from the city, so it's possible the atmospherics were not great at the time.

There are, in all, six components to Beta Scorpii and the entire system lies some 400 light years away.

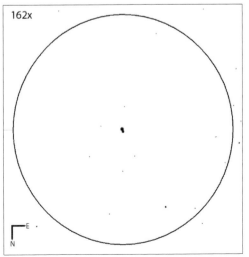

Eyepiece view courtesy *Sky Tools 3*

Jabbah

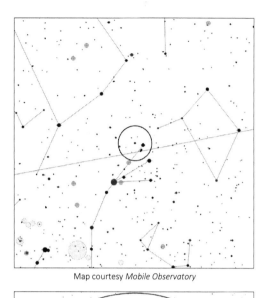

Map courtesy *Mobile Observatory*

Designation(s):	Nu Scorpii
Constellation:	Scorpius
R.A.:	16h 12m 00s
Declination:	-19° 27' 39"
Object Type:	Multiple Star
Location:	★ ★
Rating:	★ ★ ★
Best Seen:	Spring and Summer

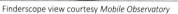

Finderscope view courtesy *Mobile Observatory*

If you're able to locate Graffias, you should have no difficulty spotting Jabbah as it lies to the east and within the same finderscope field of view.

Unfortunately, despite what you might think (or hope) Jabbah has nothing to do with *Star Wars* - but it's still a pretty interesting sight for telescope observers at both low and medium magnification.

At about 30x you'll see a pair of white stars of similar brightness at one corner of a triangle of three. But increase the magnification and what do you see?

At 81x one of the stars appeared bluish, with hints of violet but increasing the magnification again to 107x reveals a third companion.

Look carefully at the fainter component and you might see that it's split into two white stars of almost equal magnitude.

Like many other multiple stars, there are other components to this system that can't be seen with amateur equipment. In fact, it's thought there may be up to seven stars in this system, which would make for an amazing (and possibly blinding) sight if you could ever get there. And you thought a twin sunset over the deserts of Tatooine was cool!

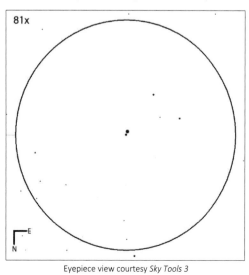

Eyepiece view courtesy *Sky Tools 3*

Designation(s):	Messier 80
Constellation:	Scorpius
R.A.:	16h 17m 02s
Declination:	-22° 58' 34"
Object Type:	Globular Cluster
Location:	★
Rating:	★
Best Seen:	Spring and Summer

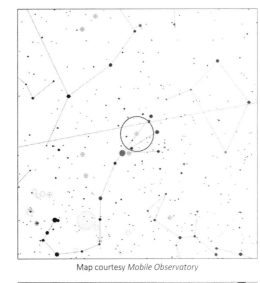

Map courtesy *Mobile Observatory*

Messier 80 is the *other* globular cluster that's easily found within the constellation of Scorpius, the Scorpion. It's often overlooked in favor of it's bigger and brighter neighbor, Messier 4, but that's not to say it's not worth seeking out.

From the suburbs, I was able to easily find the cluster at 35x. It appeared as a grey smudge close to a brighter star to the west and a group of stars to the east.

Although it's smaller than Messier 4, it appeared a little brighter to me with a bright core. Unfortunately, at both 70x and 182x there didn't seem to be much change and I wasn't able to resolve the cluster into individual stars.

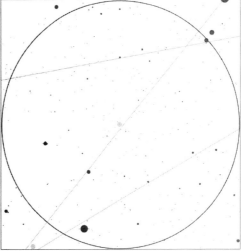

Finderscope view courtesy *Mobile Observatory*

It was another ten years before I tried again, but this time it was from the brighter skies of the city and I had less success than before. I was able to spot it at 27x, when it appeared as a very small, faint, circular grey sphere close to a star, but again, I wasn't able to improve the view.

So why hunt it down at all? Well, think about what you're seeing here. Messier 80 is over 32,000 light years away and is thought to contain about 200,000 stars, all crammed into an area some 95 light years across. In comparison, there are only about 500 stars within 100 light years of the Earth, which means it contains about 400 times as many stars as our local stellar neighborhood.

Are you still a little disappointed?

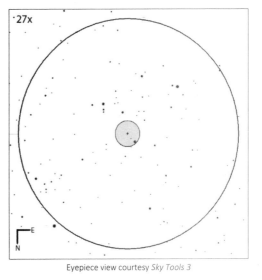

Eyepiece view courtesy *Sky Tools 3*

Messier 4

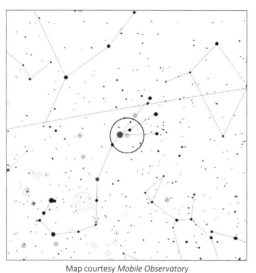

Map courtesy *Mobile Observatory*

Designation(s):	Messier 4
Constellation:	Scorpius
R.A.:	16h 23m 35s
Declination:	-26° 31' 33"
Object Type:	Globular Cluster
Location:	★
Rating:	★ ★
Best Seen:	Spring and Summer

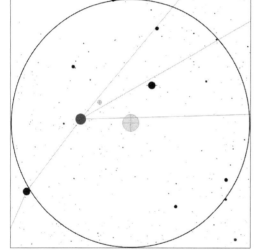

Finderscope view courtesy *Mobile Observatory*

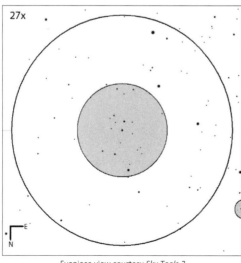

27x

Eyepiece view courtesy *Sky Tools 3*

If Messier 80 failed to impress, try your hand at Messier 4. It appears very close to Antares, the reddish star that marks the beating heart of Scorpius the Scorpion.

In fact, it's fair to say the cluster is almost unmissable – unless you're trying to find it with binoculars in the city. I've never had any luck there, but if you live under dark skies you shouldn't have much problem at all.

As with almost every deep sky object (such as star clusters, nebulae and galaxies) you'll see more under darker skies and Messier 4 is no exception.

I've been able to pick it out at low power from both the suburbs and the city, but a magnification of about 30x is only enough to show it as a very faint and fuzzy sphere.

At around 50x the cluster appeared elongated, with a band of bright stars stretching across the center and some resolution around the edges with averted vision.

City dwellers may not be so lucky but I've noted some resolution at 81x – again, with averted vision.

In comparison, at around 80x or 90x the view from suburban skies improves quite considerably and the cluster appears quite spectacular. I've noted that thousands of stars could be seen with averted vision and the cluster looked like sugar or salt spilt on black satin.

Designation(s):	Messier 6
Constellation:	Scorpius
R.A.:	17h 40m 21s
Declination:	-32° 15' 15"
Object Type:	Open Cluster
Location:	★ ★
Rating:	★ ★ ★
Best Seen:	Summer

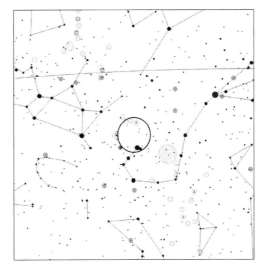

Map courtesy *Mobile Observatory*

The Butterfly cluster is one of two (the other being Messier 7) that won't be visible from the United Kingdom and skims the horizon from North America. If you live in the United States, you shouldn't have much problem as long as the whole of Scorpius can be seen from your location.

To find it, place the two stars representing the sting of the Scorpion's tail on the edge of your field of view in about the five o'clock position. (Messier 7 may be just outside the view at about the nine o'clock position.) The Butterfly cluster may appear near to the eleven o'clock position. I've noted that it's barely visible from suburban skies and is the middle of the three circles in the finderscope depiction.

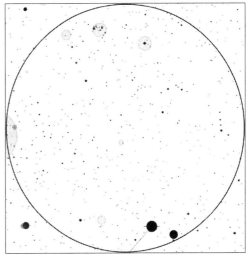

Finderscope view courtesy *Mobile Observatory*

Through binoculars it's about half the size of Messier 7 and looks like tiny, very fine granules of star dust in a triangular shape that points toward the east.

Denser and more compact than Messier 7, the cluster provides a beautiful view at 35x with its butterfly shape being quite apparent. The brightest five stars form a shape like a flattened Pleiades. Four of those are blue-white with the fifth, on the south-western side, being white, possibly orange-ish.

The majority of the stars appear much fainter, look like diamond dust and must number in the hundreds.

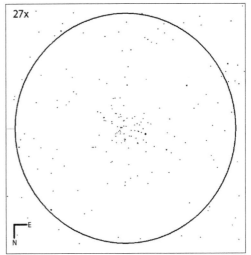

Eyepiece view courtesy *Sky Tools 3*

Messier 7

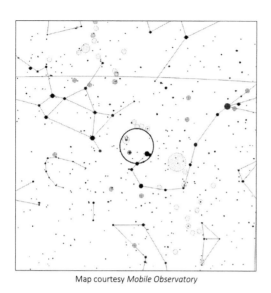

Map courtesy *Mobile Observatory*

Designation(s):	Messier 7
Constellation:	Scorpius
R.A.:	17h 53m 51s
Declination:	-34° 47' 34"
Object Type:	Open Cluster
Location:	★ ★
Rating:	★ ★
Best Seen:	Summer

Finderscope view courtesy *Mobile Observatory*

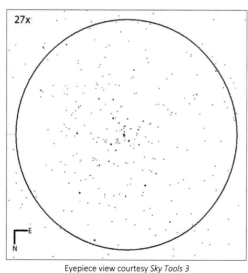

Eyepiece view courtesy *Sky Tools 3*

Close to the Butterfly Cluster in the sky, Messier 7 is larger, brighter and is consequently a little easier to see. You may be able to spot it with just your eyes under a clear, dark sky.

With the two stars of the scorpion's stinger at roughly the four o'clock position in your finder, it should be quite apparent on the opposite edge of the field of view. In fact, the two stars will point you right to it.

A low power magnification is all you'll need to enjoy this cluster. It's not as dense as the Butterfly and, in my opinion, is not as attractive. At 35x I've noted plenty of sparsely scattered stars that appear to form the shape of a bow tie (bow ties are cool, by the way.)

You may also see some gold stars among the many blue-white members; in particular, look out for an apparent double at the very heart of the cluster. Both stars are of equal brightness but with one being pale gold and the other blue-white.

Increasing the magnification to 54x may also reveal a small trio of stars forming an equilateral triangle on the eastern edge.

Messier 7 lies about 980 light years away and is thought to be about 25 light years in diameter.

The Keystone Cluster

Designation(s):	Messier 13
Constellation:	Hercules
R.A.:	16h 41m 41s
Declination:	+36° 27' 37"
Object Type:	Globular Cluster
Location:	★ ★
Rating:	★ ★ ★
Best Seen:	Spring and Summer

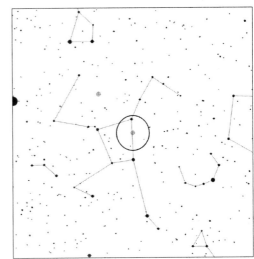

Map courtesy *Mobile Observatory*

The Keystone Cluster is by far the best globular cluster visible from the northern hemisphere.

If you have sharp eyesight and you live under clear, dark skies, you may be able to spot the cluster with just your eyes, but (more likely) you'll probably need binoculars to see it.

Through my small 8x30's from suburbia it appeared very small and looked like a faint, fuzzy star close to two other stars. It had a star-like core and appeared comet-like.

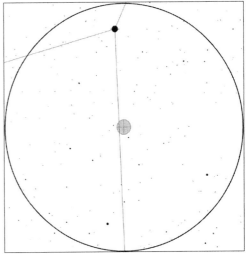

Finderscope view courtesy *Mobile Observatory*

The darker your sky, the more you'll see and you can certainly reveal a lot of detail under the right conditions. Unfortunately, it's a disappointment from the city and a shadow of its true self. Just 27x will allow you to see the cluster but it's not much better than using binoculars and increasing the magnification doesn't improve the view.

From the suburbs it's another story. At 35x I noted the bright core extending some way toward the edge, where it then faded rapidly. I also noted some resolution and – best of all – chains of stars extending out from the core.

I've found a magnification of about 100x to be best for small scopes. I've often thought it like some strange sea creature, with tentacles reaching out from the depths of space.

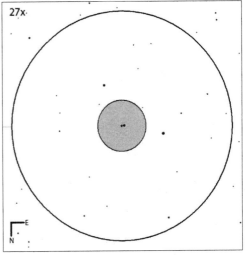

Eyepiece view courtesy *Sky Tools 3*

Rho Herculis

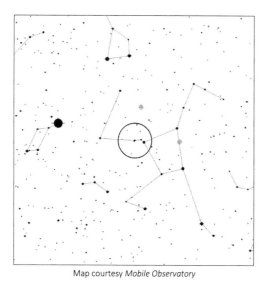

Map courtesy *Mobile Observatory*

Designation(s):	Rho Herculis
Constellation:	Hercules
R.A.:	17h 24m 12s
Declination:	+37° 08' 16"
Object Type:	Multiple Star
Location:	★ ★
Rating:	★ ★
Best Seen:	Spring and Summer

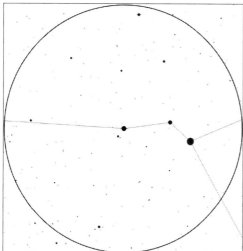

Finderscope view courtesy *Mobile Observatory*

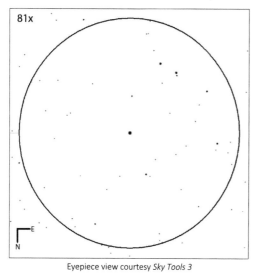

Eyepiece view courtesy *Sky Tools 3*

Rho Herculi is a white star with a nearby blue companion that's easily seen with binoculars. You might also see a fainter, third companion close to that secondary.

A telescope won't reveal much else until you get to about 50x when the primary can be barely split. Both components appear white with the primary being slightly brighter than the secondary.

Increasing the power helps to reveal some color in the pair. A magnification of about 100x shows a clear split with the primary being a pale gold. The secondary has shown some interesting colors – I've noted olive green with flecks of mustard at 107x.

The system lies just over 400 light years away and comprises a subgiant primary and a dwarf secondary. The system is thought to be about 300 million years old with a third, very faint suspected companion appearing some distance away.

Rho marks the shoulder of Hercules, the Hero. In ancient mythology, he was known as Heracles and was the mortal son of Zeus. He's famous for performing twelve labors, by order of King Eurystheus; his first was to slay a lion and his second was to slay a sea serpent. Both these creatures can now be found toward the west in the night sky, some way from Hercules, in the form of the constellations Leo and Hydra.

Kuma

Designation(s):	Nu Draconis
Constellation:	Draco
R.A.:	17h 32m 11s
Declination:	+55° 11' 03"
Object Type:	Multiple Star
Location:	★ ★ ★
Rating:	★ ★ ★
Best Seen:	Spring and Summer

Map courtesy *Mobile Observatory*

Kuma is a gem and a very easy target for small telescopes in the northern hemisphere. I was able to split the pair with both my finderscope and my small 8x30 binoculars (although I had to hold them steadily to do it. Leaning against a wall will definitely help!)

This is a double that's best observed at low power. It's an easy and clean split at just 27x with both stars appearing to be brilliant white and of equal brightness. They've always made me think of a pair of eyes staring back at me from space – in fact, some observers have taken to calling them "the eyes of the dragon" for that reason.

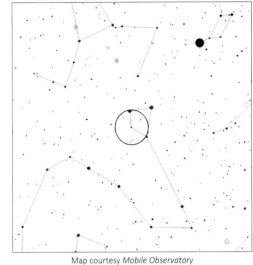

Finderscope view courtesy *Mobile Observatory*

A true multiple star system, the pair lie just under a hundred light years away with one component, Nu Draconis[2], having its own very close companion.

The constellation of Draco, the Dragon, snakes around Polaris, the Pole Star, and is actually circumpolar. This means the constellation never sets when seen from most of the northern hemisphere but appears to spin about the pole star instead.

Its head appears close to Hercules, the Hero and appears to chase him about the pole. It's at its highest in the Summer when the constellation can be seen snaking over much of the northern horizon.

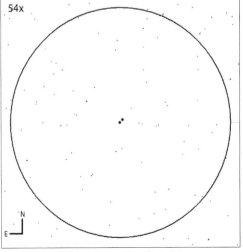

Eyepiece view courtesy *Sky Tools 3*

Messier 22

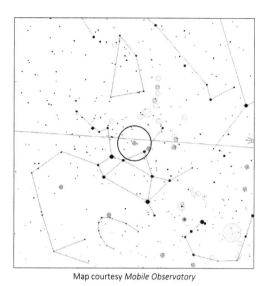

Map courtesy *Mobile Observatory*

Designation(s):	Messier 22
Constellation:	Sagittarius
R.A.:	18h 36m 24s
Declination:	-23° 54' 17"
Object Type:	Globular Cluster
Location:	★ ★
Rating:	★ ★
Best Seen:	Summer

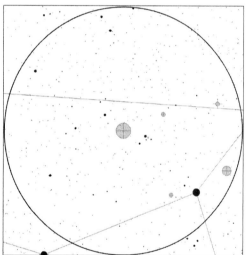

Finderscope view courtesy *Mobile Observatory*

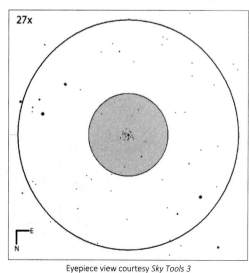

Eyepiece view courtesy *Sky Tools 3*

Messier 22, like the Keystone Cluster, is one of the best globulars in the entire night sky but unlike its cousin in Hercules, it's not well seen from the northern hemisphere.

This is a shame as it would certainly would be much better known if it were more easily seen. However, if you live in the United States it shouldn't be too much of a problem and even observers in the United Kingdom and Canada may still stand a chance.

It's conveniently located close to Kaus Borealis, the star that marks the top of the famous teapot of Sagittarius. It's easily seen with binoculars from suburban skies and appears as a small, nebulous sphere with no bright center and a uniform grey color.

A low magnification of 35x won't improve the view too much; it appears circular and could be easily mistaken for a nebula.

Increasing the magnification to 50x or higher and using averted vision might help reveal some resolution but take a look at its shape; does it still appear circular or has its appearance altered slightly? Does it look a little elongated to you?

Messier 22 is only 10,600 light years away (quite close for a globular) and contains about 70,000 stars.

The Double Double

Designation(s):	Epsilon Lyrae
Constellation:	Lyra
R.A.:	18h 44m 20s
Declination:	+39° 40' 12"
Object Type:	Multiple Star
Location:	★ ★ ★
Rating:	★ ★ ★
Best Seen:	Summer

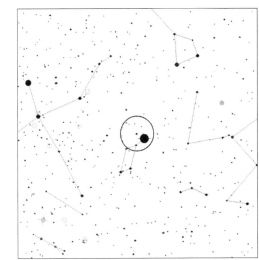

Map courtesy *Mobile Observatory*

Epsilon Lyrae, commonly known as the "Double Double," is one of the most famous and popular multiple stars in the sky – and for good reason.

First of all, turn a pair of binoculars toward it – what do you see? You might need to steady yourself against a wall, but you should easily be able to split the star into two close white components of equal brightness. It's a nice view with both Vega and Zeta Lyrae appearing within the same field of view.

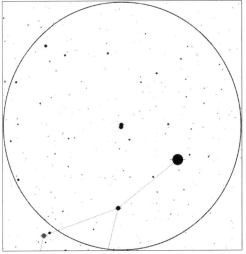

Finderscope view courtesy *Mobile Observatory*

In a small scope at low power (say, about 30x) you might notice something odd. I've seen both stars appear elongated, as though they were almost but not quite split.

Try increasing the magnification – first to about 50x, then about 75x. Are the stars split yet? Try again at about 100x and you should be able to barely split both stars into their components and the view is quite fascinating. Two pairs of bright white stars, with all four being about the same brightness. One pair might appear slightly brighter and wider than the other.

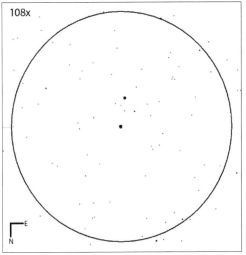

Eyepiece view courtesy *Sky Tools 3*

You're looking at a fascinating multiple star system some 162 light years away. One pair orbit each other once every 1,800 years while the other pair circle one another every 700 years. While quadruple stars are not unheard of, what's truly astonishing is that there may be as many as ten stars in this system!

Sheliak

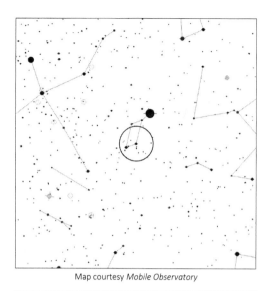

Map courtesy *Mobile Observatory*

Designation(s):	Beta Lyrae
Constellation:	Lyra
R.A.:	18h 50m 05s
Declination:	+33° 21' 46"
Object Type:	Multiple Star
Location:	★ ★ ★
Rating:	★ ★
Best Seen:	Summer

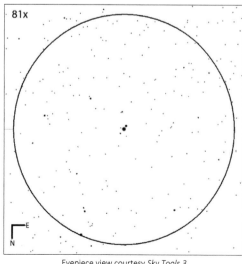

Finderscope view courtesy *Mobile Observatory*

Eyepiece view courtesy *Sky Tools 3*

Sheliak is a relatively wide double just beyond the range of regular binoculars. However, a small scope shouldn't have any problem resolving the pair at low magnification. The primary is pure white and about five or six times brighter than the blue companion.

Look carefully and you'll see the pair just off-center of a small triangle of faint stars. Increasing the magnification to around 100x will definitely make this easier to see.

Sheliak is one of those stars that makes you wish interstellar travel were possible. If you could visit this star, you'd see something quite amazing.

It lies about 950 light years away and its primary component is actually a very close double. The smaller of the two was once the giant star of the system but as it grew in size, its companion began to pull matter away from it.

In other words, the larger star was cannibalized by its smaller companion until the roles were reversed. The companion is now the dominant member of the pair and is surrounded by a disk of matter stripped away from the former giant. The two stars now orbit one another almost every thirteen days causing the star to dim slightly as seen from Earth. Try comparing its brightness to nearby stars and you may notice it fade and brighten over the intervening nights.

The Ring Nebula

Designation(s):	Messier 57
Constellation:	Lyra
R.A.:	18h 53m 35s
Declination:	+33° 01' 45"
Object Type:	Planetary Nebula
Location:	★
Rating:	★ ★ ★
Best Seen:	Summer

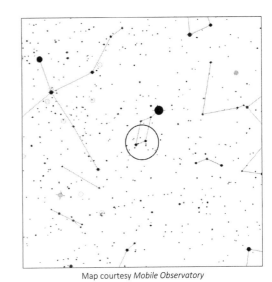

Map courtesy *Mobile Observatory*

A favorite of mine ever since I first saw it, the Ring Nebula is very appropriately named because it truly looks like a smoke ring in space.

Surprisingly, I've been able to see it from both the city and also the suburbs. At 27x it appeared circular, dark grey, small and clearly not a star. However, you won't be able to see it as a ring until you increase the magnification.

At 70x I was barely able to see the central hole and the ring-shape, but I had to use averted vision to spot it. The shape becomes more apparent as you continue to increase the magnification but you'll need to always use averted vision to see the central hole.

Finderscope view courtesy *Mobile Observatory*

At 161x I noted that it still appeared circular and cloud-like but that the nebula was well defined and the hole very apparent with averted vision. Focusing proved to be difficult while looking directly at it.

What you're seeing here is a dying star, some 2,000 light years away. The Ring Nebula is probably the best example of a planetary nebula – so-called because they look like planets when observed telescopically. It's actually a shell of gas thrown off from a giant star as it shrinks down to a white dwarf. You can actually see this star, in the center of the nebula, but you'll need a larger telescope to spot it.

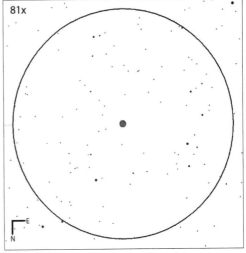

Eyepiece view courtesy *Sky Tools 3*

The Wild Duck Cluster

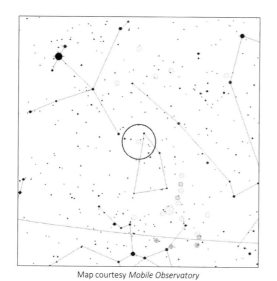

Map courtesy *Mobile Observatory*

Designation(s):	Messier 11
Constellation:	Scutum
R.A.:	18h 51m 06s
Declination:	-06° 16' 12"
Object Type:	Open Cluster
Location:	★ ★
Rating:	★ ★
Best Seen:	Summer

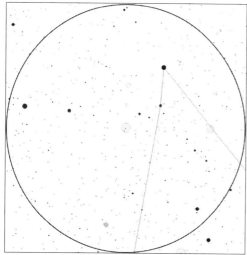

Finderscope view courtesy *Mobile Observatory*

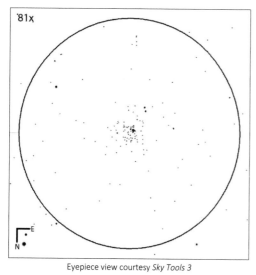

Eyepiece view courtesy *Sky Tools 3*

The Wild Duck Cluster is a small open star cluster found just to the south-west of Aquila, the Eagle. It can be easily seen with binoculars and can be found by following a slightly curved line of stars from Delta Aquilae, at the bottom of the constellation. (Alternatively, if you can easily locate Beta Scutum in your finderscope you should be able to spot the cluster within the same field of view.)

The cluster is disappointing from the city as it's small, faint and difficult to see at low power. You'll definitely need to use averted vision to really see anything of it.

If, however, you live away from the city lights, you're in for a treat. Through binoculars it appears small, compact, faint and globular – cloud-like, in fact – with a star-like point on the eastern edge.

Through a small scope at 35x you'll see a small, granular V shape with a single bright star at the tip. It's not hard to see how the cluster got its name as it looks like ducks flying south for the winter. You'll also see a few other bright stars appear within the same field of view, most notably a nearby double pair.

It's one of those clusters that definitely improves as you increase the magnification. Many more stars appear at just 54x with so many stars becoming apparent that it starts to lose its shape. The stars themselves are blue and are almost all of a uniform brightness.

The Coathanger

Designation(s):	Collinder 399
Constellation:	Vulpecula
R.A.:	19h 26m 05s
Declination:	+20° 13' 14"
Object Type:	Asterism
Location:	★ ★
Rating:	★ ★
Best Seen:	Summer

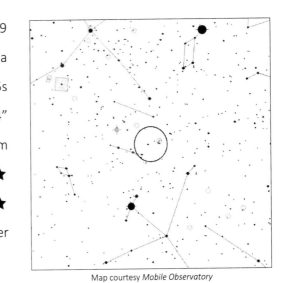

Map courtesy *Mobile Observatory*

The Coathanger is a large group of stars that's actually best observed with a very low power. In fact, you may prefer to observe it with either a pair of binoculars or even your telescope's finder.

To find it, you'll first need to locate the constellation of Sagitta, the Arrow, located a little under halfway between Albireo in Cygnus and Altair in Aquila. Look for the two stars that mark the feathers and place them at about the seven o'clock position in your field of view. The group should then be around the two o'clock mark.

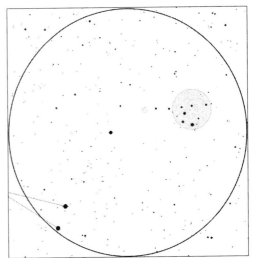

Finderscope view courtesy *Mobile Observatory*

The shape of the group should be quite apparent (albeit upside-down) through the finder or binoculars. When you turn your telescope toward it, make sure you're using your lowest power eyepiece first as you'll have difficulty fitting it into the field of view at even 35x.

The coathanger is formed by a chance alignment of eight or nine stars with the brightest forming the hook of the hanger. Technically then, this apparent "cluster" of stars is actually an asterism, meaning that it only appears to form a familiar shape by chance.

The group has been known since antiquity, having first been recorded by the Persian astronomer al-Sufi around the year 964. The cluster has also gone by the name Brocchi's Cluster, named for D. F. Brocchi, a stellar cartographer who mapped the group in the 1920's.

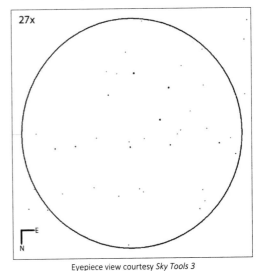

Eyepiece view courtesy *Sky Tools 3*

The Dumbbell Nebula

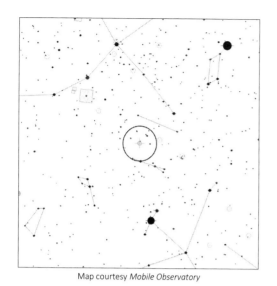

Map courtesy *Mobile Observatory*

Designation(s):	Messier 27
Constellation:	Vulpecula
R.A.:	19h 59m 36s
Declination:	+22° 43' 16"
Object Type:	Planetary Nebula
Location:	★
Rating:	★ ★
Best Seen:	Summer

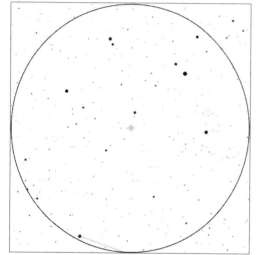

Finderscope view courtesy *Mobile Observatory*

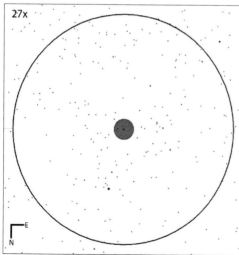

Eyepiece view courtesy *Sky Tools 3*

The Dumbbell Nebula, although faint, is actually relatively easy to find once you know where to look. You'll need to first find the constellation of Sagitta, the Arrow, about midway between Altair in Aquila and Albireo in Cygnus.

Look for Gamma Sagittae, one of the brighter stars that forms the body of the arrow itself. By placing it just on the edge or outside your finder's field of view (in the six o'clock position) you should be able to see a line of three stars near the center.

The Dumbbell Nebula is just to the south of the middle star, 14 Vulpeculae. I haven't seen it in a finder, but I've definitely snagged it with 8x30 binoculars under suburban skies. It was barely seen, circular, very misty and needed averted vision to be spotted.

If you're observing from the city, you'll probably need a telescope to see it at all. At 27x, it was seen but was extremely faint and needed averted vision. It appeared as a large, uniformly grey disk with no details seen. Increasing the magnification made observation easier.

It's much more apparent from suburban skies and should be easily seen at just 35x. It appeared rectangular at first but actually resembled a bat when observed with averted vision. Unlike the city, increasing the magnification will definitely improve the view and allow more details to be seen.

Albireo

Designation(s):	Beta Cygni
Constellation:	Cygnus
R.A.:	19h 30m 43s
Declination:	+27° 57' 35"
Object Type:	Multiple Star
Location:	★ ★ ★
Rating:	★ ★ ★
Best Seen:	Summer

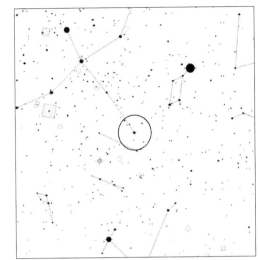

Map courtesy *Mobile Observatory*

Albireo is, by far, the most beautiful and popular double star in the entire night sky. This might sound like hyperbole, but it's really not. I challenge anyone to find a double that is more easily found, easily split and with such a striking color combination.

It's not split with a regular pair of binoculars (say 10x50's) or a finderscope, but you won't need much power to reveal both components.

It's a very easy and fairly wide split at 27x with the primary being a pale gold and about 1½ times brighter than the pale blue secondary. I've noted that the strength of the colors will sometimes vary from night to night, so it's worth taking another look from time to time.

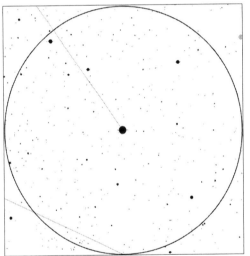

Finderscope view courtesy *Mobile Observatory*

You'll want to come back anyway – again and again – as it's a sight that's unparalleled throughout the rest of the sky. You'll show your family, your friends and your neighbors during summer barbecues and you'll anticipate their gasp – just as you did – when you first laid eyes on this beauty.

The pair lie about 430 light years away but it's not known for sure if the two components truly orbit one another. The primary, however, is itself a double but the stars are too close to be resolvable with amateur equipment.

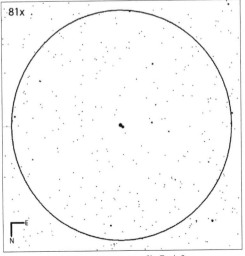

Eyepiece view courtesy *Sky Tools 3*

Messier 29

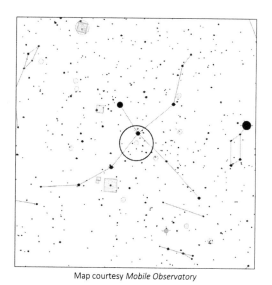

Map courtesy *Mobile Observatory*

Designation(s):	Messier 29
Constellation:	Cygnus
R.A.:	20h 23m 58s
Declination:	+38° 30' 28"
Object Type:	Open Cluster
Location:	★
Rating:	★ ★
Best Seen:	Summer

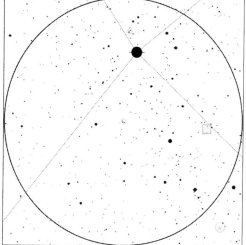

Finderscope view courtesy *Mobile Observatory*

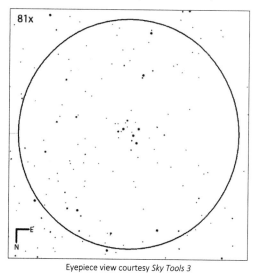

Eyepiece view courtesy *Sky Tools 3*

Everyone should have a favorite deep sky object. Whether it's a double star, an open or globular star cluster, a nebula or a galaxy, most observers have an object they feel a particular fondness for.

Mine is Messier 29. I can't explain why. It's not particularly large or spectacular and it doesn't contain hundreds of stars. It doesn't look like diamond dust on black satin. It doesn't have any golden orange stars nestled within it.

So why bother? Well, for one thing, it's easily found as Sadr, the central star in Cygnus the Swan, is a convenient location marker for it. If you place that star about halfway between the center and the edge of your finder's view (as depicted) you should have the cluster in the center.

You may not be able to see it in the finder and I've been unable to positively identify it with binoculars, but it should be readily apparent with a small telescope and a low magnification.

My first thought, upon "discovering" this cluster, was that it looked like a mini Pleiades, which came as a surprise. It's small, is surrounded by stars and has seven or eight bright stars, all white and about the same brightness.

Given its small size, it easily fits into the field of view, even at high magnification, but I've found it best at around 100x.

Zeta Sagittae

Designation(s):	Zeta Sagittae
Constellation:	Sagitta
R.A.:	19h 48m 59s
Declination:	+19° 08' 31"
Object Type:	Multiple Star
Location:	★ ★
Rating:	★ ★
Best Seen:	Summer

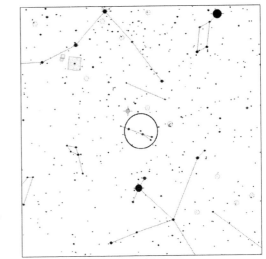

Map courtesy *Mobile Observatory*

Zeta Sagittae is a challenging double but provides a nice view for those who are able to split it. You'll need a higher power eyepiece and maybe even a Barlow for this one.

Part of the challenge may also be in locating the star to begin with. It shouldn't be much of a problem if you live in the suburbs or under a dark sky as Sagitta isn't a difficult constellation to find. Just look midway between Albireo and Altair, the brightest star in Aquila the Eagle for the small, arrow shaped constellation.

If you live in a city, you'll need to use binoculars (or scan the area with your finderscope) because the chances are the stars will be too faint to see without optical aid.

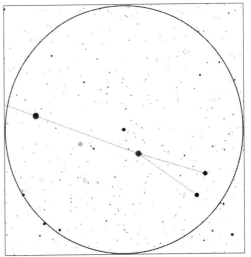

Finderscope view courtesy *Mobile Observatory*

Zeta can be seen just a little to the north of the main constellation itself and close to Delta Sagittae. You won't be able to split it with low power; 35x didn't do anything for me. At 54x, the secondary was just barely visible, but unless you've already spotted it at a higher magnification, it might elude you.

Try a magnification of about 100x and then work your way down. I was able to split the pair at 91x and saw a white primary and a tiny, faint blue secondary. Now drop the power down to a lower magnification and look for the secondary again. Keep doing this until the secondary disappears – how low can you go?

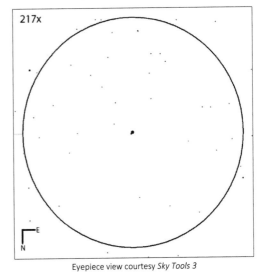

Eyepiece view courtesy *Sky Tools 3*

Messier 71

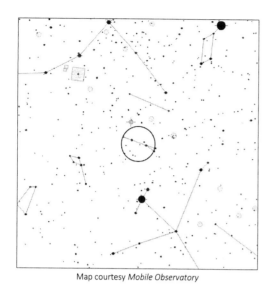

Map courtesy *Mobile Observatory*

Designation(s):	Messier 71
Constellation:	Sagitta
R.A.:	19h 53m 46s
Declination:	+18° 46' 45"
Object Type:	Globular Cluster
Location:	★
Rating:	★
Best Seen:	Summer

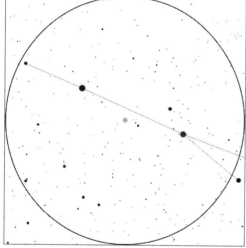

Finderscope view courtesy *Mobile Observatory*

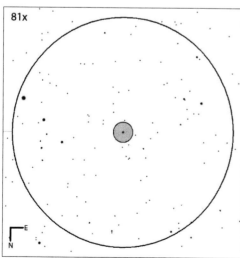

Eyepiece view courtesy *Sky Tools 3*

While you're in the vicinity of Sagitta, try your hand at Messier 71, another challenging sight for small scopes and located on just the other side of the constellation from Zeta.

It's too small and faint to be seen with in a finder but if you center the constellation in your field of view, the cluster should be visible through your telescope at low power.

This is a good example of an object you'd hunt down for the challenge, rather than for the eventual view. At 35x, it appears very faint, very small and very fuzzy. I've noted that it doesn't really look globular at all, but rather more like a nebula instead.

In fact, it doesn't even appear spherical. When I first observed it, I noted that it appeared rectangular and was uniformly dim with no bright core. Increasing the magnification to 70x improved the view a little but there was no resolution of the individual stars. (On the plus side, there's plenty of other stars in the field of view that will make it easy for you to focus.)

Messier 71 was discovered by Phillipe Loys de Chéseaux in 1746. The cluster lies about 27,000 light years away, is some 27 light years in diameter and was originally thought to be an open cluster.

Designation(s):	Alpha Capricorni
Constellation:	Capricornus
R.A.:	20h 18m 03s
Declination:	-12° 32' 41"
Object Type:	Multiple Star
Location:	★ ★ ★
Rating:	★ ★
Best Seen:	Summer

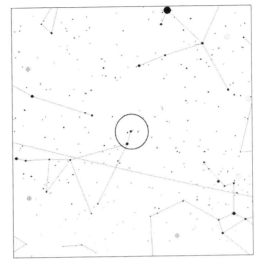

Map courtesy *Mobile Observatory*

Al Giedi is a wide, optical double, which means the two stars are not a true multiple star system.

It's a very easy object for binoculars or a finderscope with the western star appearing slightly fainter and with a hint of pale orange or cream color.

Look through a telescope and the view improves somewhat. At 35x each star will reveal a faint companion. The western star, Alpha Capricorni[2], has a wide companion while the eastern star, Alpha Capricorni[1], also has a faint companion that appears much closer to the primary star.

(Only one of these companions may be visible from the city – you may need to increase the magnification to about 80x in order to see them both. Which do you see? The wide or the close companion?)

Al Giedi is the brightest star (or stars!) in the constellation of Capricornus, the Sea Goat. It's an ancient constellation, known to a number of cultures, with a rich mythology that goes back thousands of years.

Curiously, the constellation is almost always associated with water in some way. For example, in Greek mythology, it represents the god Pan. Part goat, the god panicked while escaping the monster Typhon and was only partly successful in turning himself into a fish before hitting the water!

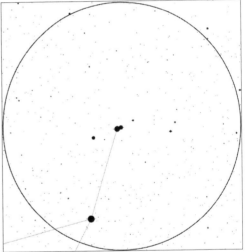

Finderscope view courtesy *Mobile Observatory*

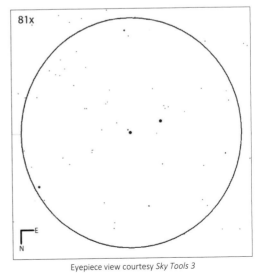

81x

Eyepiece view courtesy *Sky Tools 3*

Gamma Delphini

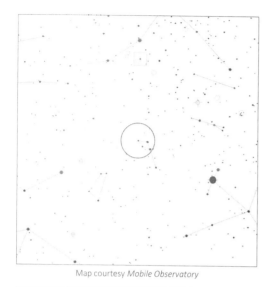

Designation(s):	Gamma Delphini
Constellation:	Delphinus
R.A.:	20h 46m 39s
Declination:	+16° 07' 38"
Object Type:	Multiple Star
Location:	★ ★ ★
Rating:	★ ★
Best Seen:	Summer

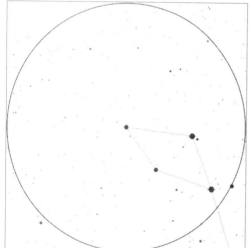

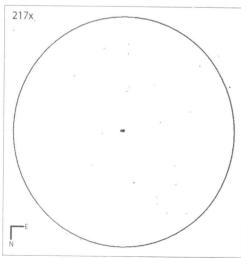

217x

Gamma Delphini is a reasonably bright star that marks one corner of a star pattern (or *asterism*) known as Job's Coffin. It's also known to be a paler version of Albireo that's an easy split for small telescopes.

Binoculars (or a finderscope) won't reveal its companion but look out for Sualocin (Alpha Delphini) just to the west and within the same field of view. You'll see a wide pair of blue-white stars with the primary about twice as bright as the secondary. (Incidentally, try spelling the name of the star backwards. It's the name of an astronomer who named it after himself!)

Gamma is easily split at 35x and is a good example of your eyes showing you color that isn't necessarily there. As explained in the notes for Lambda Arietis, two stars may appear to be of slightly different colors as a result of the brightness contrast between them.

Take a look and note their colors and then increase the magnification. Do the colors change? Maybe they will and maybe they won't. Come back on the next clear night and try it again.

I usually see a very pale yellow and very pale blue pair of stars of almost equal brightness and increasing the magnification seems to make the colors stronger. But on other nights, one looks to be pale yellow while the other is clearly white. What do you see?

Designation(s):	Messier 15
Constellation:	Pegasus
R.A.:	21h 29m 58s
Declination:	+12° 10' 01"
Object Type:	Open Cluster
Location:	★
Rating:	★ ★
Best Seen:	Summer and Autumn

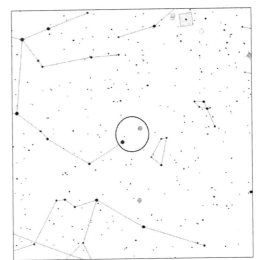

Map courtesy *Mobile Observatory*

Messier 15 is easily found, thanks to its close proximity to Enif, the brightest star in the constellation of Pegasus, the Flying Horse.

Visible in binoculars, it can be found by drawing a line through Baham (to the south-east of Enif) and Enif itself. Look for a faint, fuzzy star next to a slightly brighter one. It's easy to mistake it for a star but careful observation will reveal its non-stellar nature.

In a finder, with Enif in about the eight o'clock position, the cluster may appear at around the two o'clock mark. Again, look out for the star that appears close to it.

Unsurprisingly, the cluster is better observed through a telescope. At low power (around 30x) you'll see a small, hazy "star" with four real stars within the same field of view. The stars formed a zig-zag pattern and the cluster appeared between two of the stars. Averted vision may help to make the cluster appear a little larger.

The view is improved at a higher magnification. Between 80x and 100x I noticed that the core appeared surprisingly bright and extended about two thirds of the way toward the edge of the cluster. At 107x I some resolution was hinted at and the edges seemed to extend a little further out.

At about 12 billion years old, this cluster is one of the most ancient known.

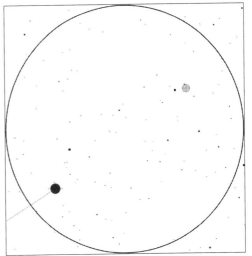

Finderscope view courtesy *Mobile Observatory*

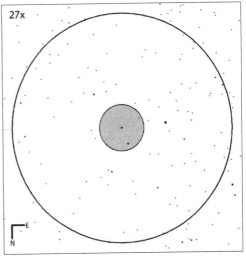

Eyepiece view courtesy *Sky Tools 3*

Observation Logs

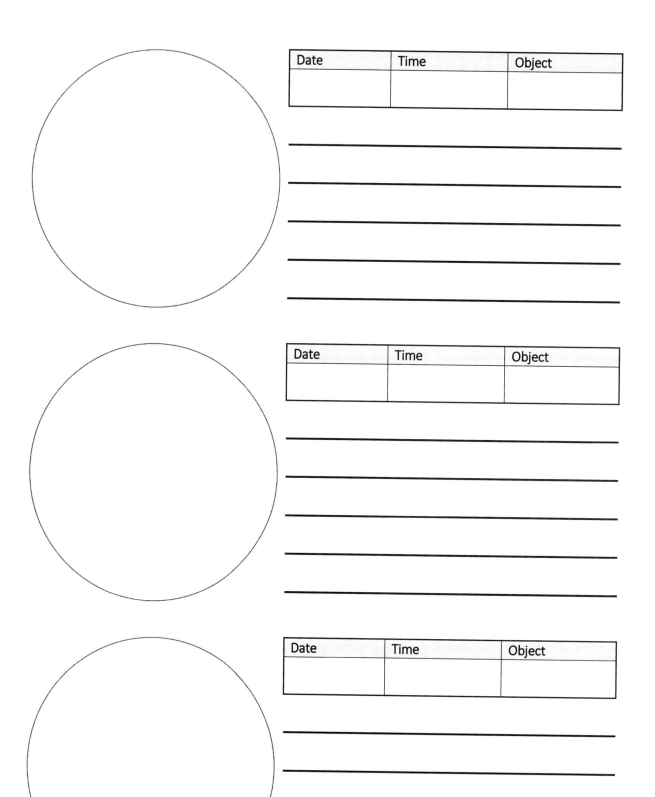

Date	Time	Object

_____

_____

_____

_____

_____

Date	Time	Object

_____

_____

_____

_____

_____

Date	Time	Object

_____

_____

_____

_____

_____

Date	Time	Object

_____

_____

_____

_____

_____

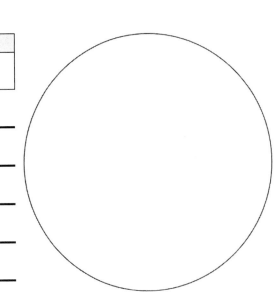

Date	Time	Object

_____

_____

_____

_____

_____

Date	Time	Object

_____

_____

_____

_____

_____

Date	Time	Object

_____

_____

_____

_____

Date	Time	Object

_____

_____

_____

_____

Date	Time	Object

_____

_____

_____

_____

Date	Time	Object

_____

_____

_____

_____

_____

Date	Time	Object

_____

_____

_____

_____

_____

Date	Time	Object

_____

_____

_____

_____

_____

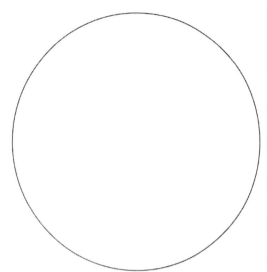

Date	Time	Object

_____

_____

_____

_____

_____

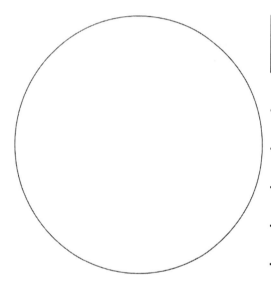

Date	Time	Object

_____

_____

_____

_____

_____

Date	Time	Object

_____

_____

_____

_____

Date	Time	Object

_____

_____

_____

_____

_____

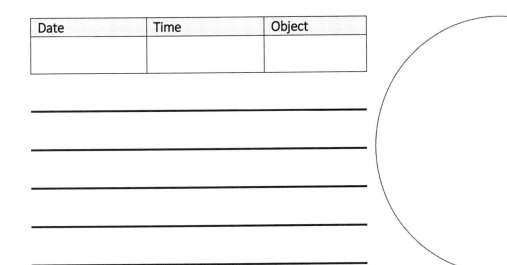

Date	Time	Object

_____

_____

_____

_____

_____

Date	Time	Object

_____

_____

_____

_____

_____

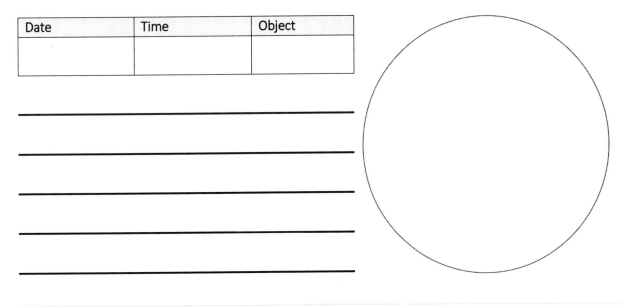

Date	Time	Object

_____

_____

_____

_____

_____

Date	Time	Object

_____

_____

_____

_____

_____

Date	Time	Object

_____

_____

_____

_____

_____

Date	Time	Object

_____

_____

_____

_____

_____

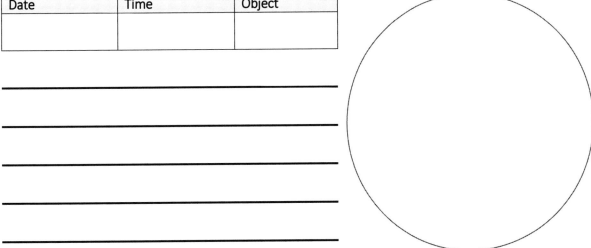

Date	Time	Object

_____

_____

_____

_____

_____

Date	Time	Object

_____

_____

_____

_____

_____

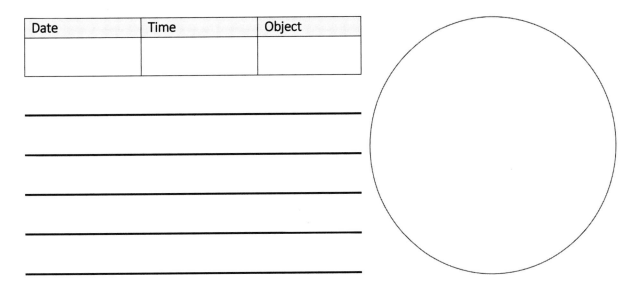

Date	Time	Object

_____

_____

_____

_____

_____

Date	Time	Object

_____

_____

_____

_____

_____

Date	Time	Object

_____

_____

_____

_____

_____

Date	Time	Object

_____

_____

_____

_____

_____

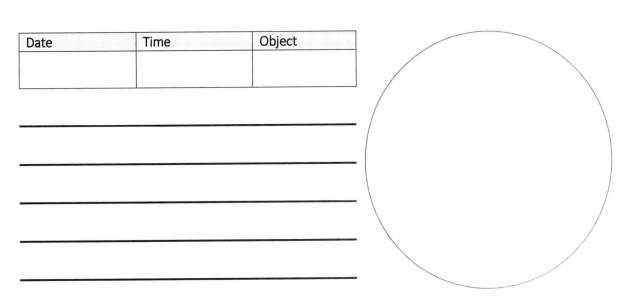

Date	Time	Object

_____

_____

_____

_____

_____

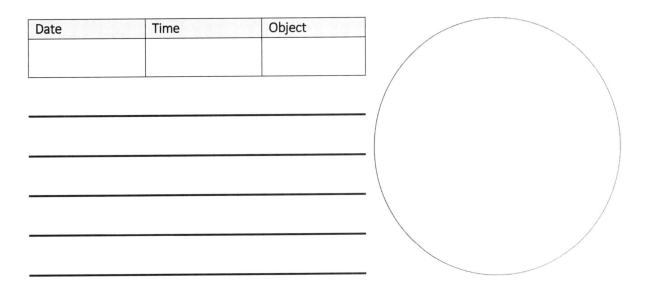

Date	Time	Object

_____

_____

_____

_____

_____

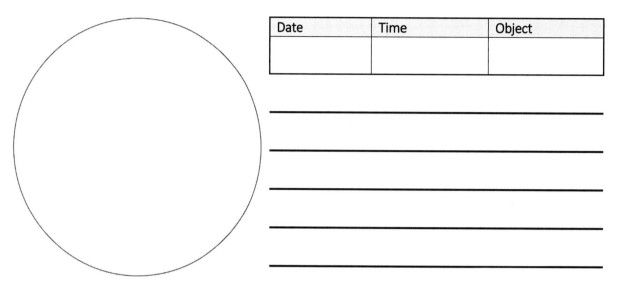

Date	Time	Object

_____

_____

_____

_____

_____

Date	Time	Object

_____

_____

_____

_____

_____

Date	Time	Object

_____

_____

_____

_____

_____

Date	Time	Object

_____

_____

_____

_____

_____

Date	Time	Object

_____

_____

_____

_____

_____

Date	Time	Object

_____

_____

_____

_____

_____

Date	Time	Object

_____

_____

_____

_____

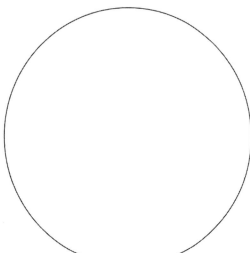

Date	Time	Object

_____

_____

_____

_____

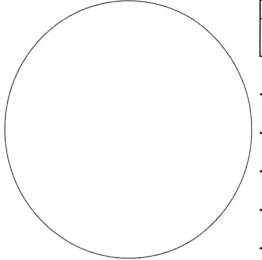

Date	Time	Object

_____

_____

_____

_____

Date	Time	Object

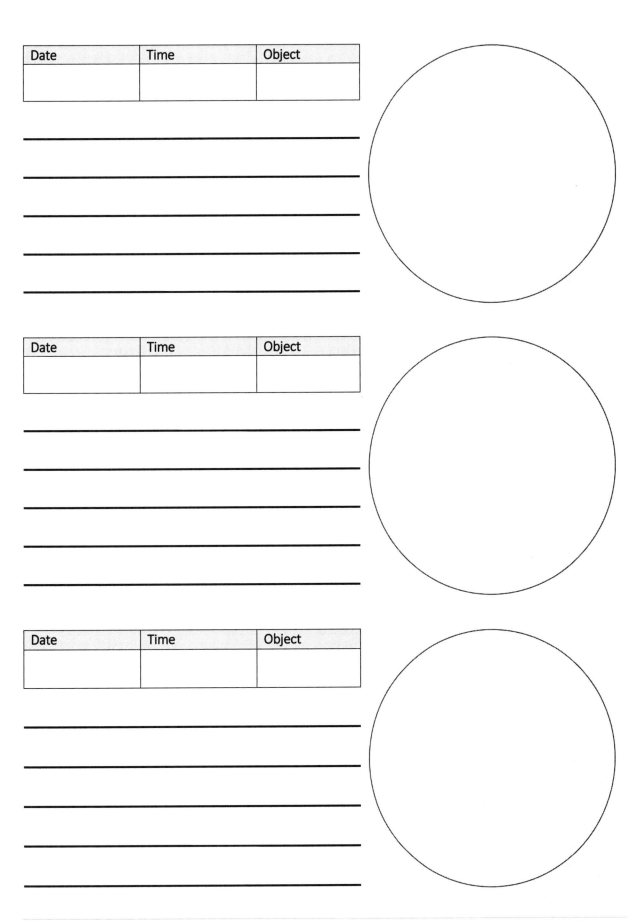

_____

_____

_____

_____

_____

Date	Time	Object

_____

_____

_____

_____

_____

Date	Time	Object

_____

_____

_____

_____

_____

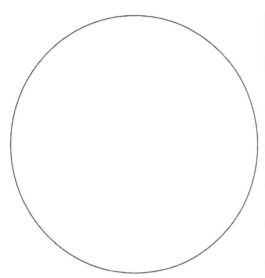

Date	Time	Object

_____

_____

_____

_____

_____

Date	Time	Object

_____

_____

_____

_____

_____

Date	Time	Object

_____

_____

_____

_____

_____

Date	Time	Object

_____

_____

_____

_____

_____

Date	Time	Object

_____

_____

_____

_____

_____

Date	Time	Object

_____

_____

_____

_____

_____

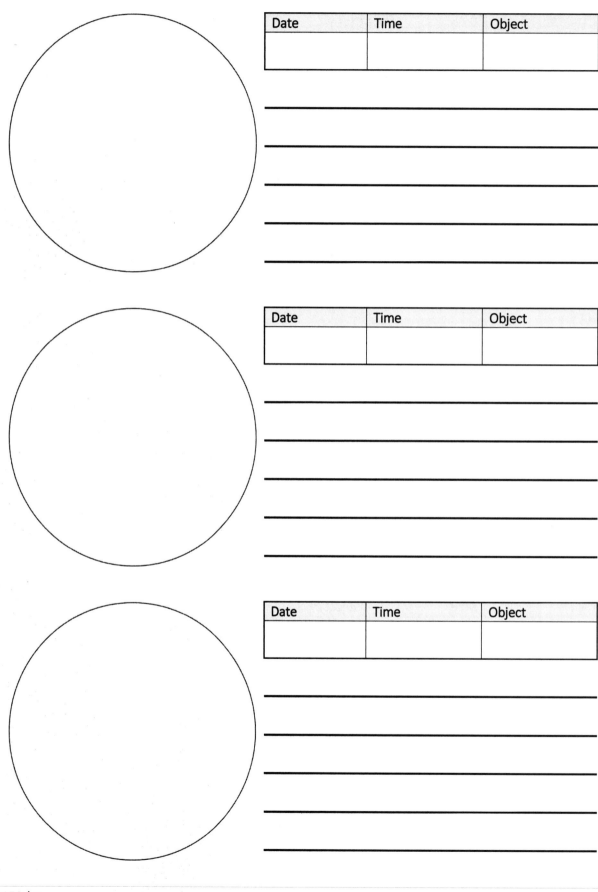

Date	Time	Object

Date	Time	Object

Date	Time	Object

Date	Time	Object

_____

_____

_____

_____

_____

Date	Time	Object

_____

_____

_____

_____

_____

Date	Time	Object

_____

_____

_____

_____

_____

Date	Time	Object

_____

_____

_____

_____

_____

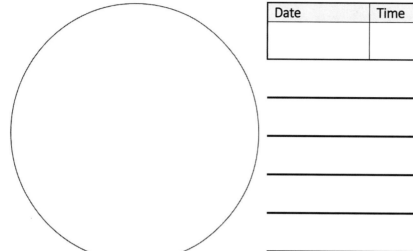

Date	Time	Object

_____

_____

_____

_____

_____

Date	Time	Object

_____

_____

_____

_____

Date	Time	Object

_____

_____

_____

_____

_____

Date	Time	Object

_____

_____

_____

_____

_____

Date	Time	Object

_____

_____

_____

_____

_____

Deep Sky Objects by Constellation

Note: The coordinates have been slightly abbreviated and rounded up to the nearest minute for formatting reasons. Inputting these coordinates into a GoTo controller (with zero as the seconds) should still allow your telescope to find the object.

ANDROMEDA

Autumn and Winter

	Designation	R.A.	Dec.	Type	Location	Rating	Page
Pi Andromedae	Pi Andromedae	00h 37m	+33° 43'	Multiple Star	★★	★★	68
Andromeda Galaxy	Messier 31	00h 43m	+41° 16'	Spiral Galaxy	★★	★★	69
NGC 752	NGC 752	01h 58m	+37° 52'	Open Cluster	★★	★★	70
Almach	Gamma Andromedae	02h 04m	+42° 20'	Multiple Star	★★★	★★★	71

ARIES

Autumn and Winter

	Designation	R.A.	Dec.	Type	Location	Rating	Page
Mesarthim	Gamma Arietis	01h 53m	+19° 18'	Multiple Star	★★★	★★★	76
Lambda Arietis	Lambda Arietis	01h 58m	+23° 36'	Multiple Star	★★★	★★★	77

AURIGA

Winter

	Designation	R.A.	Dec.	Type	Location	Rating	Page
Messier 37	Messier 37	05h 52m	+32° 33'	Open Cluster	★	★★★	88

BOÖTES

Spring and Summer

	Designation	R.A.	Dec.	Type	Location	Rating	Page
Arcturus	Alpha Boötis	14h 16m	+19° 11'	Multiple Star	★★★	★	106
Delta Boötis	Delta Boötis	15h 16m	+33° 19'	Multiple Star	★★★	★★★	107

CANCER

	Designation	R.A.	Dec.	Type	Location	Rating	Page
Praesepe	Messier 44	08h 40m	+19° 40'	Open Cluster	★★	★★★	96
Iota Cancri	Iota Cancri	08h 47m	+28° 46'	Multiple Star	★★★	★★★	97
Messier 67	Messier 67	08h 51m	+11° 49'	Open Cluster	★	★★	98

CANIS MAJOR

	Designation	R.A.	Dec.	Type	Location	Rating	Page
Sirius	Alpha Canis Majoris	06h 45m	-16° 43'	Multiple Star	★★★	★	92
Messier 41	Messier 41	06h 46m	-20° 45'	Open Cluster	★★	★★★	93

CANIS MINOR

	Designation	R.A.	Dec.	Type	Location	Rating	Page
Procyon	Alpha Canis Minoris	07h 39m	+05° 13'	Multiple Star	★★★	★	94

CANES VENATICI

	Designation	R.A.	Dec.	Type	Location	Rating	Page
Cor Caroli	Alpha Canum Venaticorum	12h 56m	+38° 19'	Multiple Star	★★★	★★★	104

CAPRICORNUS

	Designation	R.A.	Dec.	Type	Location	Rating	Page
Al Giedi	Alpha Capricorni	20h 18m	-12° 33'	Multiple Star	★★★	★★	129

CASSIOPEIA

	Designation	R.A.	Dec.	Type	Location	Rating	Page
Achird	Eta Cassiopeiae	00h 49m	+57° 49'	Multiple Star	★★★	★★★	72
Owl Cluster	NGC 457	01h 20m	+58° 17'	Open Cluster	★	★★★	73
Messier 103	Messier 103	01h 33m	+60° 39'	Open Cluster	★	★	74
NGC 663	NGC 663	01h 46m	+61° 13'	Open Cluster	★★	★★★	75

CORVUS

	Designation	R.A.	Dec.	Type	Location	Rating	Page
Algorab	Delta Corvi	12h 30m	-16° 31'	Multiple Star	★★★	★★	103

CYGNUS

	Designation	R.A.	Dec.	Type	Location	Rating	Page
Albireo	Beta Cygni	19h 31m	+27° 58'	Multiple Star	★★★	★★★	125
Messier 29	Messier 29	20h 24m	+38° 30'	Open Cluster	★	★★	126

DELPHINUS

	Designation	R.A.	Dec.	Type	Location	Rating	Page
Gamma Delphini	Gamma Delphini	20h 47m	+16° 08'	Multiple Star	★★★	★★	130

DRACO

	Designation	R.A.	Dec.	Type	Location	Rating	Page
Kuma	Nu Draconis	17h 32m	+55° 11'	Multiple Star	★★★	★★★	117

GEMINI

Winter

	Designation	R.A.	Dec.	Type	Location	Rating	Page
Messier 35	Messier 35	06h 09m	+24° 21'	Open Cluster	★★	★★★	89
Castor	Alpha Geminorum	07h 35m	+31° 53'	Multiple Star	★★★	★★★	90

HERCULES

Spring and Summer

	Designation	R.A.	Dec.	Type	Location	Rating	Page
Keystone Cluster	Messier 13	16h 42m	+36° 28'	Globular Cluster	★★	★★★	115
Rho Herculis	Rho Herculis	17h 24m	+37° 08'	Multiple Star	★★	★★	116

LEO

Spring

	Designation	R.A.	Dec.	Type	Location	Rating	Page
Regulus	Alpha Leonis	10h 08m	+11° 58'	Multiple Star	★★★	★	99
Adhafera	Zeta Leonis	10h 17m	+23° 25'	Multiple Star	★★★	★★	100
Algieba	Gamma Leonis	10h 20m	+19° 51'	Multiple Star	★★★	★★★	101
Denebola	Beta Leonis	11h 49m	+14° 34'	Multiple Star	★★★	★★	102

LEPUS

Winter

	Designation	R.A.	Dec.	Type	Location	Rating	Page
Gamma Leporis	Gamma Leporis	05h 44m	-22° 27'	Multiple Star	★★★	★★	87

LIBRA

Spring

	Designation	R.A.	Dec.	Type	Location	Rating	Page
Zuben Elgenubi	Alpha Librae	14h 51m	-16° 00'	Multiple Star	★★★	★★	108

LYRA

Summer

	Designation	R.A.	Dec.	Type	Location	Rating	Page
Double Double	Epsilon Lyrae	18h 44m	+39° 40'	Multiple Star	★★★	★★★	119
Sheliak	Beta Lyrae	18h 50m	+33° 22'	Multiple Star	★★★	★★	120
Ring Nebula	Messier 57	18h 54m	+33° 02'	Multiple Star	★	★★★	121

MONOCEROS

Winter

	Designation	R.A.	Dec.	Type	Location	Rating	Page
Beta Monocerotis	Beta Monocerotis	06h 29m	-07° 02'	Multiple Star	★★	★★	91

ORION

Winter

	Designation	R.A.	Dec.	Type	Location	Rating	Page
Mintaka	Delta Orionis	05h 32m	-00° 18'	Multiple Star	★★★	★★	83
Meissa	Lambda Orionis	05h 35m	+09° 56'	Multiple Star	★★★	★★★	84
Orion Nebula	Messier 42	05h 35m	-05° 23'	Nebula	★★★	★★★	85
Sigma Orionis	Sigma Orionis	05h 39m	-02° 36'	Multiple Star	★★★	★★★	86

PEGASUS

Autumn

	Designation	R.A.	Dec.	Type	Location	Rating	Page
Messier 15	Messier 15	21h 30m	+12° 10'	Globular Cluster	★	★★	131

PERSEUS

Autumn and Winter

	Designation	R.A.	Dec.	Type	Location	Rating	Page
Double Cluster	NGC 869 & 884	02h 22m	+57° 09'	Open Clusters	★★★	★★★	78
Messier 34	Messier 34	02h 42m	+42° 45'	Open Cluster	★★	★★	79

PUPPIS

Winter

	Designation	R.A.	Dec.	Type	Location	Rating	Page
Messier 93	Messier 93	07h 44m	-23° 51'	Open Cluster	★	★★	95

SAGITTA

Summer

	Designation	R.A.	Dec.	Type	Location	Rating	Page
Zeta Sagittae	Zeta Sagittae	19h 49m	+19° 09'	Multiple Star	★★	★★	127
Messier 71	Messier 71	19h 54m	+18° 47'	Globular Cluster	★	★	128

SAGITTARIUS

Summer

	Designation	R.A.	Dec.	Type	Location	Rating	Page
Messier 22	Messier 22	18h 36m	-23° 54'	Globular Cluster	★★	★★	118

SCORPIUS

Summer

	Designation	R.A.	Dec.	Type	Location	Rating	Page
Graffias	Beta Scorpii	16h 05m	-19° 48'	Multiple Star	★★★	★★★	109
Jabbah	Nu Scorpii	16h 12m	-19° 28'	Multiple Star	★★	★★★	111
Messier 80	Messier 80	16h 17m	-22° 59'	Globular Cluster	★	★	112
Messier 4	Messier 4	16h 24m	-26° 32'	Globular Cluster	★	★★	113
Butterfly Cluster	Messier 6	17h 40m	-32° 15'	Open Cluster	★★	★★★	114
Messier 7	Messier 7	17h 54m	-34° 48'	Open Cluster	★★	★★	115

SCUTUM

Summer

	Designation	R.A.	Dec.	Type	Location	Rating	Page
Wild Duck Cluster	Messier 11	18h 51m	-06° 16'	Open Cluster	★★	★★	122

TAURUS

	Designation	R.A.	Dec.	Type	Location	Rating	Page
Pleiades	Messier 45	03h 48m	+24° 10'	Open Cluster	★★★	★★★	81
Crab Nebula	Messier 1	05h 35m	+22° 01'	Supernova Remnant	★	★	82

URSA MAJOR

	Designation	R.A.	Dec.	Type	Location	Rating	Page
Mizar & Alcor	Zeta and 80 Ursae Majoris	13h 24m	+54° 56'	Multiple Star	★★★	★★★	105

URSA MINOR

	Designation	R.A.	Dec.	Type	Location	Rating	Page
Polaris	Alpha Ursae Minoris	02h 32m	+89° 16'	Multiple Star	★★★	★	80

VULPECULA

	Designation	R.A.	Dec.	Type	Location	Rating	Page
The Coathanger	Collinder 399	19h 26m	+20° 13'	Asterism	★★	★★	123
Dumbbell Nebula	Messier 27	20h 00m	+22° 43'	Planetary Nebula	★	★★	124

The Greek Alphabet

Alpha	α	Epsilon	ε	Iota	ι	Nu	ν	Rho	ρ	Phi	φ
Beta	β	Zeta	ζ	Kappa	κ	Xi	ξ	Sigma	σ	Chi	χ
Gamma	γ	Eta	η	Lambda	λ	Omicron	o	Tau	τ	Psi	ψ
Delta	δ	Theta	θ	Mu	μ	Pi	π	Upsilon	υ	Omega	ω

Elongations of Mercury and Venus, 2016-2025

	Mercury		Venus	
2016	February 6th	Morning Sky		
	April 17th	Evening Sky		
	June 4th	Morning Sky		
	August 16th	Evening Sky		
	September 28th	Morning Sky		
	December 10th	Evening Sky		
2017	January 18th	Morning Sky	January 11th	Evening Sky
	March 31st	Evening Sky	June 2nd	Morning Sky
	May 17th	Morning Sky		
	July 29th	Evening Sky		
	September 11th	Morning Sky		
	November 23rd	Evening Sky		
2018	January 1st	Morning Sky	August 16th	Evening Sky
	March 14th	Evening Sky		
	April 28th	Morning Sky		
	July 11th	Evening Sky		
	August 26th	Morning Sky		
	November 5th	Evening Sky		
	December 14th	Morning Sky		
2019	February 26th	Evening Sky	January 5th	Morning Sky
	April 11th	Morning Sky		
	June 23rd	Evening Sky		
	August 9th	Morning Sky		
	October 19th	Evening Sky		
	November 27th	Morning Sky		
2020	February 9th	Evening Sky	March 24th	Evening Sky
	March 23rd	Morning Sky	August 12th	Morning Sky
	June 3rd	Evening Sky		
	July 21st	Morning Sky		
	September 30th	Evening Sky		
	November 9th	Morning Sky		
2021	January 23rd	Evening Sky	October 29th	Evening Sky
	March 5th	Morning Sky		
	May 16th	Evening Sky		
	July 4th	Morning Sky		
	September 13th	Evening Sky		
	October 24th	Morning Sky		

	Mercury			Venus	
2022	January 6th	Evening Sky		March 19th	Morning Sky
	February 16th	Morning Sky			
	April 28th	Evening Sky			
	June 15th	Morning Sky			
	August 26th	Evening Sky			
	October 8th	Morning Sky			
	December 20th	Evening Sky			
2023	January 29th	Morning Sky		June 3rd	Evening Sky
	April 11th	Evening Sky		October 23rd	Morning Sky
	May 28th	Morning Sky			
	August 9th	Evening Sky			
	September 21st	Morning Sky			
	December 3rd	Evening Sky			
2024	January 11th	Morning Sky			
	March 24th	Evening Sky			
	May 9th	Morning Sky			
	July 21st	Evening Sky			
	September 4th	Morning Sky			
	November 15th	Evening Sky			
	December 24th	Morning Sky			
2025	March 7th	Evening Sky		January 9th	Evening Sky
	April 20th	Morning Sky		May 31st	Morning Sky
	July 3rd	Evening Sky			
	August 18th	Morning Sky			
	October 29th	Evening Sky			
	December 7th	Morning Sky			

Oppositions of Mars, Jupiter and Saturn, 2016-2025

	Mars		Jupiter		Saturn	
2016	May 21st	Scorpius	March 7th	Leo	June 2nd	Ophiuchus
2017			April 7th	Virgo	June 14th	Ophiuchus
2018	July 26th	Capricornus	May 8th	Libra	June 27th	Sagittarius
2019			June 9th	Ophiuchus	July 8th	Sagittarius
2020	October 13th	Pisces	July 14th	Sagittarius	July 20th	Sagittarius
2021			August 19th	Capricornus	August 1st	Capricornus
2022	December 7th	Taurus	September 26th	Pisces	August 13th	Capricornus
2023			November 2nd	Aries	August 27th	Aquarius
2024			December 7th	Taurus	September 7th	Aquarius
2025	January 15th	Gemini			September 20th	Pisces

Complete List of Constellations

Latin	Genitive	English Name	Abbreviation	Size
Andromeda	Andromedae	The Princess	And	19th
Antila	Antilae	The Air Pump	Ant	62nd
Apus	Apodis	The Bird of Paradise	Aps	67th
Aquarius	Aquarii	The Water Bearer	Aqr	10th
Aquila	Aquilae	The Eagle	Aql	22nd
Ara	Arae	The Altar	Ara	63rd
Aries	Arietis	The Ram	Ari	39th
Auriga	Aurigae	The Charioteer	Aur	21st
Boötes	Boötis	The Herdsman	Boo	13th
Caelum	Caeli	The Chisel	Cae	81st
Camelopardalis	Camelopardalis	The Giraffe	Cam	18th
Cancer	Cancri	The Crab	Cnc	31st
Canes Venatici	Canum Venaticorum	The Hunting Dogs	CVn	38th
Canis Major	Canis Majoris	The Large Dog	CMa	43rd
Canis Minor	Canis Minoris	The Little Dog	CMi	71st
Capricornus	Capricorni	The Sea Goat	Cap	40th
Carina	Carinae	The Keel	Car	34th
Cassiopeia	Cassiopeiae	The Queen	Cas	25th
Centaurus	Centauri	The Centaur	Cen	9th
Cepheus	Cephei	The King	Cep	27th
Cetus	Ceti	The Sea Monster	Cet	4th
Chamaeleon	Chamaeleontis	The Chameleon	Cha	79th
Circinus	Circini	The Compass	Cir	85th
Columba	Columbae	The Dove	Col	54th
Coma Berenices	Comae Berenices	Berenices' Hair	Com	42nd
Corona Australis	Coronae Australis	The Southern Crown	CrA	80th
Corona Borealis	Coronae Borealis	The Northern Crown	CrB	73rd
Corvus	Corvi	The Crow	Crv	70th
Crater	Crateris	The Cup	Crt	53rd
Crux	Crucis	The Southern Cross	Cru	88th
Cygnus	Cygni	The Swan	Cyg	16th
Delphinus	Delphini	The Dolphin	Del	69th
Dorado	Doradus	The Goldfish	Dor	72nd
Draco	Draconis	The Dragon	Dra	8th
Equuleus	Equulei	The Foal	Equ	87th
Eridanus	Eridani	The River	Eri	6th

Complete List of Constellations (cont.)

Latin	Genitive	English Name	Abbreviation	Size
Fornax	Fornacis	The Furnace	For	41st
Gemini	Geminorum	The Twins	Gem	30th
Grus	Gruis	The Crane	Gru	45th
Hercules	Herculis	The Hero	Her	5th
Horologium	Horologii	The Clock	Hor	58th
Hydra	Hydrae	The Water Snake	Hya	1st
Hydrus	Hydri	The Lesser Water Snake	Hyi	61st
Indus	Indi	The Indian	Ind	49th
Lacerta	Lacertae	The Lizard	Lac	68th
Leo	Leonis	The Lion	Leo	12th
Leo Minor	Leonis Minoris	The Little Lion	LMi	64th
Lepus	Leporis	The Hare	Lep	51st
Libra	Librae	The Scales	Lib	29th
Lupus	Lupi	The Wolf	Lup	46th
Lynx	Lyncis	The Lynx	Lyn	28th
Lyra	Lyrae	The Lyre	Lyr	52nd
Mensa	Mensae	The Table Mountain	Men	75th
Microscopium	Microscopii	The Microscope	Mic	66th
Monoceros	Monocerotis	The Unicorn	Mon	35th
Musca	Muscae	The Fly	Mus	77th
Norma	Normae	The Carpenter's Level	Nor	74th
Octans	Octantis	The Octant	Oct	50th
Ophiuchus	Ophiuchi	The Serpent Bearer	Oph	11th
Orion	Orionis	The Hunter	Ori	26th
Pavo	Pavonis	The Peacock	Pav	44th
Pegasus	Pegasi	The Flying Horse	Peg	7th
Perseus	Persei	The Hero	Per	24th
Phoenix	Phoenicis	The Phoenix	Phe	37th
Pictor	Pictoris	The Painter's Easel	Pic	59th
Pisces	Piscium	The Fishes	Psc	14th
Piscis Austrinus	Piscis Austrini	The Southern Fish	PsA	60th
Puppis	Puppis	The Poop Deck	Pup	20th
Pyxis	Pyxidis	The Compass	Pyx	65th
Reticulum	Reticuli	The Net	Ret	82nd

Latin	Genitive	English Name	Abbreviation	Size
Sagitta	Sagittae	The Arrow	Sge	86th
Sagittarius	Sagittarii	The Archer	Sgr	15th
Scorpius	Scorpii	The Scorpion	Sco	33rd
Sculptor	Sculptoris	The Sculptor	Scl	36th
Scutum	Scuti	The Shield	Sct	84th
Serpens	Serpentis	The Serpent	Ser	23rd
Sextans	Sextantis	The Sextant	Sex	47th
Taurus	Tauri	The Bull	Tau	17th
Telescopium	Telescopii	The Telescope	Tel	57th
Triangulum	Trianguli	The Triangle	Tri	78th
Triangulum Australe	Trianguli Australis	The Southern Triangle	TrA	83rd
Tucana	Tucanae	The Toucan	Tuc	48th
Ursa Major	Ursae Majoris	The Great Bear	UMa	3rd
Ursa Minor	Ursae Minoris	The Little Bear	UMi	56th
Vela	Velorum	The Sails	Vel	32nd
Virgo	Virginis	The Virgin	Vir	2nd
Volans	Volantis	The Flying Fish	Vol	76th
Vulpecula	Vulpeculae	The Fox	Vul	55th

Glossary

Aperture

The aperture of a telescope is the width of the lens on a refractor telescope or the width of the primary mirror of a reflector telescope tube. In both cases, this is the end that points up toward the sky.

See also *Reflector Telescope* and *Refractor Telescope*

Asterism

An asterism is a chance alignment of stars that gives the impression of a recognizable shape or pattern in the sky. For example, the seven brightest stars of Ursa Major form an asterism known as the Big Dipper in North America or the Plough in northern Europe. Similarly, the brightest stars of Sagittarius form a shape known as the Teapot.

Astronomical Unit

One Astronomical Unit (AU) is the distance from the Earth to the Sun. This is taken as 149,597,870 kilometers or 92,955,806 miles. To make life (and polite conversation as your astronomical society) easier, this is taken as 150 million kilometers or 93 million miles.

See also *Light Year*

Averted Vision

Averted vision is the "trick" of looking at an object out of the corner of your eye. As your eyes are more sensitive to light (rather than color) with averted vision, you're more likely to see a particularly faint object. Averted vision can be particularly useful in observing the details in a nebula.

Barlow Lens

A Barlow lens is an accessory that allows you to effectively double (or, with some, triple) the magnification of any eyepiece used with it. A Barlow is considered by many to be an essential tool for the amateur astronomer.

Declination

Just as geographical coordinates use longitude and latitude, stellar cartographers use right ascension and declination, respectively. Declination, like latitude, is measured in degrees and indicates an object's position in the sky relative to the northern and southern hemisphere. For example, Polaris has a declination of almost +90° and is therefore almost overhead from the North Pole. An object with a declination of 0° will pass overhead when observed from the equator. There are no bright stars with a declination of -90° and, therefore, no star marking the celestial South Pole.

See also *R.A. (Right Ascension)*

Focal Length

Focal length is the distance that light travels from the telescope's entry point (i.e., the lens on a refractor telescope or the primary mirror on a reflector) to the eyepiece. It is measured in millimeters. The longer the focal length, the higher the possible magnification of the telescope.

See also *Aperture, Reflector Telescope* and *Refractor Telescope.*

Globular Clusters

There are two types of star cluster: open and globular. Globular star clusters are like huge balls of stars, with each one containing thousands of stars. These stars are packed into a relatively small space, often with only a few light years between each one. The clusters form a halo around the galactic center and mostly contain older stars, often billions of years old.

See also *Open Clusters*

Greatest Eastern/Western Elongation

As both Mercury and Venus orbit the Sun at a closer distance than the Earth, they never stray very far from the Sun in the sky. However, there comes a time when each planet has moved the furthest it can before it starts to move back towards the Sun again. This point is called its greatest elongation.

Greatest eastern elongation means the planet is at its furthest point to the east of the Sun, thereby making it visible in the evening sky (confusingly, in the *west* after sunset.) When the planet is at its greatest western elongation, it's visible in the morning sky in the *east* before sunrise.

See also *Opposition*

Light Year

A light year is the distance that light travels in a year. Light travels at an approximate speed of 186,000 miles per *second*. In other words, at a distance of roughly 92 million miles, light from the Sun takes about eight minutes to reach us.

Given that the distance from the Earth to the Sun is one *Astronomical Unit*, it can be said that light travels 7 ½ AUs every hour, 180 AUs every day, 1,260 AUs every week, about 5,400 every month and 65,520 every year. The closest star is Proxima Centauri, which, at a distance of 4.24 light years, is some 277,804 AUs away. (And this is why we don't measure every distance in miles and kilometers.)

See also *Astronomical Unit*

Nebula (pl. nebulae)

A nebula is a huge cloud of gas and dust in space. These clouds are light years in diameter and are the birthplaces of the stars themselves. As the gas and dust particles move through space, they collide with other particles and gradually begin to grow in mass. Eventually, the clump of gas becomes so massive that a gravitational collapse begins, causing nuclear fusion and the birth of a star. Most of the nebulae are located away from bright stars (and may require a little practice to locate) but the Orion Nebula is easily seen with unaided eyes on a winter's night.

At the other end of a star's life are planetary nebulae. A planetary nebula is the last dying gasp of a star. It's the shell of gas and dust that the star ejects as it collapses and dies, causing the shell to expand out into space. They're named *planetary* nebulae because many look like the tiny disc of a planet when observed through a telescope.

If the star goes supernova, you may well be able to see the remnants in the night sky. Many of these are faint (such as the Veil Nebula, which actually looks like the remains of an exploded star) but one, the Crab Nebula, should also be visible on a winter's night.

Open Clusters

Open clusters are a type of star cluster. Whereas globular clusters appear small, spherical and often faint, most open clusters are quite the opposite. Many contain tens or hundreds of stars of varying brightness, scattered across an area of sky sometimes larger than the full Moon. There are a number that are easily observed in a small telescope, with the Pleiades, Praesepe and Butterfly clusters being fine examples.

See also *Globular Clusters*

Opposition

A planet (or other solar system body, such as an asteroid) is said to be at opposition when it's directly opposite the Sun in the sky. When this occurs, it rises at sunset and sets at sunrise and is therefore visible throughout the entire night. This is when the planet is at its best for observation, as it will be at its closest to the Earth and will appear largest and brightest in the night sky.

Opposition can only occur if the body orbits the Sun at a greater distance than the Earth. For example, the planets Mars, Jupiter and Saturn all periodically reach opposition but the planets Mercury and Venus will always appear close to the Sun in the sky.

See also *Greatest Eastern/Western Elongation*

Optical Binary

An optical binary is a chance alignment of two stars in the sky, thereby giving the illusion that the pair are a true multiple star system. In reality, the stars may be hundreds of light years apart and moving in different directions through space.

See also *Primary Component* and *Secondary Component*

Planetary Nebulae

See *Nebula*

Primary Component

The primary component of a multiple star system is the brightest with the next brightest being the secondary. For example, the famous double star Albireo has a golden primary with a sapphire blue secondary.

See also *Optical Binary* and *Secondary component*

R.A. (Right Ascension)

On star maps and charts, R.A. is the celestial equivalent of longitude and enables astronomers to easily locate an object in the night sky. Unlike longitude, it's measured in hours, minutes and seconds with 00h 00m 00s holding a special significance. When the Sun reaches that point in its movement across the sky, the March (vernal) equinox occurs. Spring then begins in the northern hemisphere while autumn begins in the south.

See also *Declination*

Reflector Telescope

A reflector telescope is one that works using mirrors. The light enters through the open end (the *aperture*) and then hits the primary mirror at the bottom of the telescope tube. The light is then reflected back up towards the aperture, where it hits a smaller secondary mirror and is directed into the eyepiece. This has the benefit of having the eyepiece conveniently located near the top of the tube and also allows for a longer *focal length* without increasing the length of the tube itself.

See also *Aperture, Focal Length* and *Refractor Telescope.*

Refractor Telescope

A refractor telescope uses lenses to collect and focus light for the observer. It comprises a long tube with a primary lens at the aperture, where the light enters the telescope. The light then travels all the way down the tube and exits through the eyepiece at the bottom.

See also *Aperture, Focal Length* and *Reflector Telescope.*

Secondary Component

The secondary component of a multiple star system is typically the second brightest star, with the brightest being the primary.

See also *Optical Binary* and *Primary Component*

Supernova Remnant

See *Nebula.*

About the Author

Photo by James Bartlett

A former monthly columnist for Astronomy magazine, Richard J. Bartlett has had a passion for the stars since the age of six.

His first website, StarLore, was featured in Sky & Telescope magazine in early 2001. Following that success, he freelanced for Astronomy, reviewing astronomical websites and software in his monthly Webweaver Picks column.

Additionally, he has moderated on the Universe Today forums, operated his own astronomical message board and still manages his own space news website, AstroNews. (http://astronewsus.wordpress.com/).

His latest blog, The Astronomical Year, highlights current astronomical events on an almost daily basis. (http://theastronomicalyear.wordpress.com/)

Now living in the suburbs of Los Angeles, he still stops to stare at the sometimes smoggy night sky through the city's light-pollution.

Also by the Author...

Amazon US: http://tinyurl.com/rjbamazon-us   *Amazon UK*: http://tinyurl.com/rjbamazon-uk

The Amateur Astronomer's Notebook: A Journal for Recording and Sketching Astronomical Observations

With over 300 pages, this journal has enough room to record notes from 150 observing sessions.

The Astronomical Almanac: A Comprehensive Guide to Night Sky Events

This almanac details thousands of astronomical events over the next five years.

An Astronomical Year

A perennial quick-reference guide gives astronomers the information they need for all the year's upcoming events.

The Deep Sky Observer's Guide

Designed for the astronomer looking to discover something new, this handy guide lists suggested deep sky objects to be observed on any night and at any time.

The Easy Guide to the Night Sky

Written for the amateur astronomer who wants to discover more in the night sky, this book explores the constellations and reveals many of the highlights visible with just your eyes or binoculars.

The Night Sky Sights

Aimed at the absolute beginner, this annual guide highlights best astronomical events to be seen with just your eyes over the coming year.

The Wonder of It All

From our home here on Earth out to the stars beyond our own solar system, this journey will take you and your child on an adventure to discover our unique place within the universe.

The Author and Orion Online

Orion Telescopes & Binoculars
Website: http://www.OrionTelescopes.com/
Facebook: https://www.facebook.com/oriontelescopes/
Twitter: https://twitter.com/oriontelescopes

Richard J. Bartlett
Website: http://tinyurl.com/theastroyear
Facebook: http://tinyurl.com/theastroyear
Twitter: http://tinyurl.com/rjbtwitter
Email: astronomywriter@gmail.com

Other Recommended Resources

Books

365 Starry Nights – Chet Raymo

Astronomy Hacks – Robert Bruce Thompson & Barbara Fritchman Thompson

Backyard Astronomer's Guide – Terence Dickinson & Alan Dyer

Binocular Highlights – Gary Seronik

Celestial Harvest – James Mullaney

Celestial Sampler – Sue French

Cosmic Challenge – Philip S. Harrington

Deep Sky Observing with Small Telescopes – David J. Eicher

Deep-Sky Wonders – Sue French

Deep Sky Wonders – Walter Scott Houston

Double Stars for Small Telescopes – Sissy Haas

Illustrated Guide to Astronomical Wonders – Robert Bruce Thompson & Barbara Fritchman Thompson

Observer's Sky Atlas, The – E. Karkoschka

Planet Observer's Handbook, The – Fred W. Price

Pocket Sky Atlas – Roger W. Sinnott

Star-Hopping for Backyard Astronomers – Alan M. MacRobert

Star Names – Their Lore and Meaning – Richard Hinckley Allen

Star Watch – Philip S. Harrington

Touring the Universe Through Binoculars – Phillip S. Harrington

Turn Left at Orion – Guy Consolmagno and Dan M. Davis

Software

Mobile Observatory by Wolfgang Zima (http://zima.co)

Orion StarSeek (Mobile App – http://www.OrionTelescopes.com/StarSeek)

Sky Tools by Greg Crinklaw (http://www.skyhound.com)

Stellarium (Open Source software – http://www.stellarium.org)